Making Mathematics Accessible for Elementary Students Who Struggle

Using CRA/CSA for Interventions

Making Mathematics Accessible for Elementary Students Who Struggle

Using CRA/CSA for Interventions

Margaret Flores, PhD, BCBA-D
Megan Burton, PhD
Vanessa Hinton, PhD

5521 Ruffin Road
San Diego, CA 92123

e-mail: info@pluralpublishing.com
Website: http://www.pluralpublishing.com

Copyright © 2018 by Plural Publishing, Inc.

Typeset in 11/13 Garamond by Flanagan's Publishing Services, Inc.
Printed in the United States of America by McNaughton & Gunn

All rights, including that of translation, reserved. No part of this publication may be reproduced, stored in a retrieval system, or transmitted in any form or by any means, electronic, mechanical, recording, or otherwise, including photocopying, recording, taping, Web distribution, or information storage and retrieval systems without the prior written consent of the publisher.

For permission to use material from this text, contact us by
Telephone: (866) 758-7251
Fax: (888) 758-7255
e-mail: permissions@pluralpublishing.com

Every attempt has been made to contact the copyright holders for material originally printed in another source. If any have been inadvertently overlooked, the publishers will gladly make the necessary arrangements at the first opportunity.

Library of Congress Cataloging-in-Publication Data

Names: Flores, Margaret, 1972- author. | Burton, Megan, author. | Hinton, Vanessa, author.
Title: Making mathematics accessible for elementary students who struggle : using CRA/CSA interventions / Margaret Flores, PhD, BCBA-D, Megan Burton, PhD, Vanessa Hinton, PhD.
Description: San Diego, CA : Plural Publishing, [2018] | Includes bibliographical references and index.
Identifiers: LCCN 2017024348| ISBN 9781597569842 (alk. paper) | ISBN 1597569844 (alk. paper)
Subjects: LCSH: Mathematics--Study and teaching. | Cognition in children. | Learning disabled children--Education--Mathematics.
Classification: LCC QA135.6 .F5945 2018 | DDC 371.92--dc23
LC record available at https://lccn.loc.gov/2017024348

Contents

Preface — *vii*

CHAPTER 1
Purpose — 1

CHAPTER 2
Number Sense and the
Concrete-Representational/Semi-Concrete–Abstract Sequence — 11

CHAPTER 3
Counting, Cardinality, and the
Concrete-Representational/Semi-Concrete–Abstract Sequence — 29

CHAPTER 4
Teaching Addition Using the
Concrete-Representational/Semi-Concrete–Abstract Sequence — 59

CHAPTER 5
Teaching Subtraction Using the
Concrete-Representational/Semi-Concrete–Abstract Sequence — 95

CHAPTER 6
Teaching Multiplication Using the
Concrete-Representational/Semi-Concrete–Abstract Sequence — 127

CHAPTER 7
Teaching Division Using the
Concrete-Representational/Semi-Concrete–Abstract Sequence — 167

CHAPTER 8
Understanding Fractions Using the
Concrete-Representational/Semi-Concrete–Abstract Sequence — 183

CHAPTER 9
Operations With Fractions Using the
Concrete-Representational/Semi-Concrete–Abstract Sequence — 225

Index — *267*

Preface

The three authors of this book are former teachers and current university faculty who prepare teachers and conduct research in the area of mathematics instruction for students with and without disabilities. Our areas of expertise are different since we come from elementary education and special education, but our purpose in writing this book is to provide teachers with information about learning trajectories and strategies for teaching students who struggle in mathematics. There is much overlap of ideas across our fields, but differences in terminology and approaches may sometimes overshadow our common purposes. This book was born out of many conversations, collaborative activities, and research among the authors about effective mathematics interventions. Students who struggle may have disabilities or may be students who lack prerequisite understanding and skills. These students' intervention teachers may have general education backgrounds or special education backgrounds, and both need information about intervention practices from general education and special education. It is the authors' intent to provide teachers with approaches and methods that are of sufficient intensity to support students but also guide students toward independence. Each chapter shows how the concrete-representational/semi-concrete–abstract (CRA/CSA) sequence is used to support student learning across elementary mathematics concepts. The use of objects (concrete level), pictures, and drawings (representational/semi-concrete) prior to instruction using numbers only (abstract) will provide students with experiences that build conceptual understanding and then lead students to procedural knowledge and fluency. It is important that students are able to explain their thinking and reasoning at each phase of this process and to see the connections between the different representations of content that are learned through CRA/CSA. Learning mathematics using CRA/CSA will allow students to develop conceptual, procedural, and declarative knowledge that will be used to engage in mathematical practices.

CHAPTER 1

Purpose

The purpose of this book is to provide teachers with methods and strategies appropriate for elementary students receiving tiered interventions within a response to intervention framework or students with disabilities in need of intensive instruction. This book was developed out of conversations and research collaborations between three university faculty with expertise in special education services at the intermediate elementary and middle school levels, early childhood and primary elementary school levels, and general education elementary in the area of mathematics. Given the rigorous demands of current mathematics standards, the three authors synthesized current special education and general education research and best practices to develop practical instructional guidelines that will allow students who struggle with mathematics to have access to grade-level standards. Currently, there are many research articles that support the use of effective intensive instruction; however, reading and implementing the content of these research articles is unrealistic for practicing or preservice teachers. Therefore, this book synthesized, interpreted, and made research findings applicable to classroom implementation.

This book was written for teachers or preservice teachers who currently or will provide small group instruction to students who have not been successful within the Tier 1 or the universal level of the response to intervention framework because they have learning gaps in prerequisite skills or they have identified or unidentified disabilities. The instructional guidelines within each chapter are intensive and provide students with repeated practice; therefore, it is likely that the methods and content are too explicit and inappropriate for whole-group grade-level instruction. It is the authors' intention to provide mathematics preservice and practicing specialists, interventionists, and special education teachers with a guide to evidence-based, intensive mathematics instructional practices that address elementary standards. The elementary standards included in this book range from number sense, numbers and operations, and fraction concepts. The content of elementary mathematics standards used across the United States is addressed, but in order for this book to be useful across regions, the authors did not specifically label the standards. The content addressed by the methods in each chapter is aligned with the content of current elementary standards, all or most of which current and preservice teachers are knowledgeable.

STUDENT CHALLENGES ADDRESSED BY THIS BOOK

Students for whom this book is intended have learning gaps in prerequisite mathematics conceptual understanding and skills and/or immature conceptions of numbers and operations. Up to 6% of students in schools have an identified mathematics learning disability, and students who display a mathematics learning disability usually have trouble performing mathematics throughout school and into adulthood (Powell et al., 2013). Many more students struggle with mathematics without a formal diagnosis of a mathematics learning disability (Powell et al.). Immature conceptions of numbers and mathematics concepts may begin as early as kindergarten; lower mathematics performance, as early as kindergarten, results in smaller gains in mathematics throughout students' schooling (Jordan et al., 2007).

Researchers have shown that students with mathematical difficulties may have difficulty in information processing, including working memory (Kroesbergen et al., 2014). Working memory makes it possible for students to build connections among number symbols, quantities, and mathematical principles (Kolkman et al., 2013). The main task of working memory is to simultaneously store, monitor, and encode incoming information, and process or activate new information (Kroesbergen et al., 2014). Numerical tasks require students to simultaneously and sequentially perceive, code, and compare information about numbers, often making connections between various representations.

Working memory is made up of a central executive system, which coordinates the processing of incoming information with three subordinate systems (Baddeley, 2000, 2010): the phonological loop, visual-spatial sketchpad, and episodic buffer. The phonological loop stores verbal information. Numerical verbal information involves encoding and processing number words and numerals, and linguistically retrieving representations of numbers and mathematical concepts from long-term memory (Mononen, Aunio, & Koponen, 2014). The visual-spatial sketchpad stores visual and spatial information. Numeric visual and spatial information involves comparisons of numbers, estimating, and mapping information on a mental number line. The episodic buffer links information to form integrated units of visual, spatial, and verbal information.

Students who have mathematical difficulties often have trouble performing tasks that involve simultaneous verbal processing and verbal storage and performing tasks that require simultaneous numerical information processing and numerical information storage (Peng & Fuchs, 2016). Students' numeric working memory retrieves information about mathematical concepts, integrates that information with incoming information, and updates the previously learned information while students perform the current mathematical task. It is through experiences of applying verbal and visual number symbols to quantities that expand one's numeric working memory and shape mathematical abilities (Kolkman et al., 2013). Instructional mathematical tasks that provide students with experience in storing and processing numbers and operations improve working memory and mathematics performance (Kroesbergen et al., 2014). Therefore, instruction for students who demonstrate difficulty with number sense needs to include integrated nonsymbolic (objects and drawings) and symbolic practice (numbers) in processing and activating knowledge about numbers and operations. An effective way of combining nonsymbolic and symbolic skill instruction for students who demonstrate numerical difficulties is through the use of the concrete-representational/semi-concrete–abstract sequence of instruction. Therefore, this book is intended to provide preservice and practicing special education teachers, mathematics specialists, and interventionists with the instructional tools to intervene and build students' conceptual and procedural understanding of mathematics by addressing standards from kindergarten through fifth grade.

WHAT IS THE CONCRETE-REPRESENTATIONAL/ SEMI-CONCRETE–ABSTRACT SEQUENCE?

Due to the abstract nature of mathematics, learning about mathematical concepts is dependent on the representation of the mathematical ideas (Lesh, Landau, & Hamilton, 1983). There are five types of representations used to understand mathematics. They include real-life experiences, manipulatives, pictures, spoken words, and written symbols (Lesh et al., 1983). These five representations fit within Bruner's (1966) three stages of how people use representations to understand information. The stages are the enactive stage, iconic stage, and symbolic representation stage. The enactive stage involves manipulative objects (counters, base-ten blocks, or other items that can be physically moved) without an internal representation of the objects, meaning that the student would not have a mental picture of that object when thinking about it. The iconic stage includes development of mental images of what has been manipulated, meaning that the students can visualize objects in their mind. The third stage is the symbolic, in which information is stored in the form of code or symbols that can be classified and organized.

The concrete-representational/semi-concrete–abstract teaching sequence is based on Bruner's (1966) stages of mental representation. This teaching sequence was first published in the research literature with students with disabilities when Miller and Mercer (1992, 1993) used the concrete-semi-concrete–abstract sequence to teach basic operations to elementary students with disabilities. At the concrete level, students learned to solve basic equations (e.g., 2 + 3 =) using objects, and the mathematical symbols were translated into words and objects were used to solve the problem. Miller and Mercer found that students with disabilities needed an average of three lessons to master an operation using objects. The representational phase involved the use of pictures or student-generated drawings to solve problems. At this phase, students were not dependent upon objects, and drawings assisted in their eventual transition to mental representation. After students mastered computation at the representational phase, they learned a strategy to assist in learning the final phase in which visual aids would not be available. The strategy involved steps to assist the student in attending to critical steps in computation such as attending to symbols and numbers; this attention to equation details helped students to make the appropriate mental visual representation before solving the problem at the abstract phase. The final phase, abstract, involved solving equations using numbers only without any external visual prompts to aid in mental representation.

There was a shift in the special education mathematics literature to use another term, *representational* (Harris, Miller, & Mercer, 1995; Miller & Morin, 1998), and it has been used more frequently and exclusively in the special education literature ever since. However, the term *CSA* was also used through the early 21st century (Jordan, Miller, & Mercer, 1998; Maccini & Hughes, 2000; Maccini & Ruhl, 2000). Currently, the special education literature has referred to this teaching sequence using objects, drawings, and numbers only as the concrete-representational–abstract sequence (CRA). It may be argued that the first descriptor, CSA, was more specific and accurate to the actual task since pictures are somewhat concrete and cannot be physically moved and touched like objects but manipulated through drawing. The use of the term *representational* in CRA is less specific since *representational* is a generic term that could be used to refer to objects and numeric symbols, which both represent numbers and not just drawings and pictures. Since both of these terms are present in the literature and educators may have a specific preference for terminology based on past history, philosophical preference, or practical use, both will be used throughout the book using the concrete-representational/semi-concrete–abstract and the abbreviation (CRA/CSA).

The CRA/CSA sequence involves the use of objects, pictures/drawings, and numbers only. Teachers may address these levels of instruction using materials that are most readily available. At the concrete level, students must be able to physically move objects that represent numbers when learning about numbers themselves or when solving equations, or representing and manipulating fractions. This may be accomplished through teacher-made objects using card stock (e.g., base 10 blocks formed with squares representing hundreds, those same squares cut into 10 equal strips of paper to represent tens, and those 10 strips cut into 10 equal squares to represent ones). This may be the least expensive, although labor-intensive approach to creating concrete objects. Using commercially available sets of base 10 blocks would be the next option that would be readily available in many elementary schools. There are also virtual manipulatives that are available on websites and applications; there is some research supporting their use (Bouck & Flanagan, 2010). This option can be used if there is adequate student access to devices that allow actual movement and manipulation of objects. **It is most critical that use of virtual manipulatives at the concrete level allow students to physically manipulate the objects**; otherwise, the student will have a representational/semi-concrete experience while missing the concrete. If virtual manipulatives are available, they are engaging for students and less messy since there are no management issues related to distribution and clean-up. The authors do not advocate one particular method if they can all be used appropriately to develop students' conceptual knowledge. The chapters in this book can all be taught using any of these materials.

HOW THIS BOOK ADDRESSES STUDENTS' CHALLENGES THROUGH CRA/CSA

In addition to the information-processing deficits described previously, students who struggle in mathematics may have difficulties in processing language. Mathematics presents new and expanded academic language to students as they progress through school. Mathematical symbols and their accompanying language present challenges to students who may already have difficulties acquiring, making meaning, and expressing their understanding of language. The CRA/CAS sequence assists students in the acquisition of mathematics concepts and the accompanying language by first providing a concrete experience with numbers and symbols. As quantities and operations are taught, students have physical objects to assist with the connections between prior knowledge and new mathematical concepts. For example, addition is joining or combining. Concrete-level instruction physically shows this process and the verbal language associated with the mathematical equation involving symbols come to life as groups of physical objects that have been placed next to the addends are placed and moved toward each other into a larger group on the right side of the equal sign. The meaning of equal is shown by discussing and showing that both sides of the equation are the same. When encoding this new information about symbols, numbers, and language, students have a nonsymbolic experience to encode and connect with prior everyday experiences in which objects are put together into a larger group.

The representational/semi-concrete level includes pictures and/or student-developed drawings. The experience is nonsymbolic but closer to the eventual symbolic understanding that is needed to progress in mathematics since physical movement is not included. Students' conceptual understanding is further developed at this stage. Receptive language, or encoding or taking in information, is aided with nonsymbolic representation, and the concept and language learned at the concrete level is reinforced with a different medium. Expressive language, or telling and showing ones' learning, is aided because the visual representation supports the student's generation of words needed to explain the process.

Finally, the abstract phase allows for instruction in the mathematical process that reflects and leads to fluent computation. It is imperative that students have, through the previous levels of instruction, an understanding and a solid mental construct of the mathematical operation prior to the abstract level of instruction. Usually, students learn a strategy just prior to abstract instruction in order to assist them with procedural components of a mathematical process. Students who have information-processing deficits also struggle in thinking about their own thinking, or metacognition. Metacognitive deficits tend to manifest through impulsively, lack of attention to detail, and poor self-monitoring. Teaching students a procedural strategy provides them with cues and prompts for attending to components of an equation or number and remembering to check one's work, monitoring their actions. This ensures that students encode accurate information with which to make meaning and complete a mathematical task as well as monitor their work just in case there was a breakdown in the encoding, retrieval, and expression processes. The CRA/CSA sequence is included in every chapter addressing instruction related to number sense, counting, basic operations, complex operations, basic fraction concepts, and operations with fractions. Instruction in one of these major concepts from the elementary standards is shown within the context of the CRA/CSA sequence to ensure that students who struggle with mathematics build firm conceptual understanding of mathematics concepts and related academic language that will build on further learning and academic success.

EXPLICIT INSTRUCTION

As this book is written for students in need of intensive interventions, instructional lessons using CRA/CSA are presented using the explicit instruction model. The authors are aware that students who develop firm mathematical understanding can question and explore without such direct assistance. However, this book is written for students who do not have the prerequisite skills and need specific directed instruction that will eventually lead to less teacher guidance and more student exploration. Most chapters include examples of thinking aloud, and this is the eventual goal for students who struggle, talking about their mathematical thinking, understanding, and problem-solving steps. Through explicit instruction using CRA/CSA, teachers provide students with the tools and the language for this skill. Explicit instruction involves five instructional steps: (a) an advance organizer that includes a review of prerequisite learning, sets the stage for learning, communicates the topic and structure of the lesson, and explains why it is relevant to the student; (b) modeling in which the teacher physically shows and thinks aloud while completing the instructional task; (c) guided practice in which the teacher and students complete the instructional task together; (d) independent practice in which students complete the instructional task without assistance and the teacher monitors and provides feedback when needed; and (e) post organizer in which the teacher discusses highlights of the instructional task with the students. This instructional model is likely more customary for special education teachers, but general education teachers may be less comfortable with this approach at first. However, explicit instruction is necessary for students who have already failed using other approaches. Students who are already behind, need to learn more in less time, potentially catching up to their peers who have mastered the target concepts and are moving on to more complex mathematics. This book is written based on the belief that after firm conceptual understanding, procedural knowledge, and basic fluency are mastered, students are ready for a less teacher-directed approach to explore more advanced mathematical concepts.

HOW THIS BOOK IS ORGANIZED AND BEST UTILIZED

This book is organized around major mathematics concepts within elementary standards and presented in order of their presentation from kindergarten to Grade 5. The standards addressed are number sense, numbers, basic operations, complex operations, fraction concepts, and fractions with operations. The authors integrated problem solving into each chapter and intentionally excluded a chapter on problem solving.

The rationale for this type of presentation of problem solving was to encourage readers to address problem solving within the context of instruction in numbers, operations, and fractions. Students who struggle with mathematics have a difficult time with generalization, and teaching real-world application at the beginning of conceptual learning will assist them in making connections between mathematics and their life experiences. In addition, infusing problem solving within the CRA/CSA process will provide opportunities to teach, connect, and make meaning of mathematics language, symbols, and everyday language that are necessary to solve word problems. In the authors' opinion, separating numbers and operations from word problems only serves to further disconnect mathematics from students' ability to make meaning.

Each chapter includes similar structure and organization with explanations of concepts and various approaches to teaching that each include the use of the CRA/CSA sequence. Figure and pictorial explanations of instruction and use of instructional materials are included. In chapters related to operations when the materials are representations of the base 10 number system, drawings are used since pictures would likely not enhance the message. Within the chapters related to fractions, pictures of objects at the concrete level are used to demonstrate how instructors should vary students' experiences with materials that show numbers made of parts that comprise a whole.

One may prefer to use this book as a guide to intervention across elementary levels, beginning with the first chapter and using chapters in order. Another approach would be to use chapters that are particularly relevant to the reader's practical experience and needs. Each chapter is written so that it could stand alone; one does not build and necessarily include prerequisite knowledge to understand the proceeding chapter.

SUMMARY

This book is written for teachers (practicing or preservice) of students who struggle in elementary mathematics, students who have failed to master grade-level standards after whole-group evidence-based universal instruction, and students with identified disabilities. The methods and procedures within this book are intended for small group intensive instruction, one that addresses students' need for increased repetition and explicitness that cannot be provided within a large group of students with diverse learning needs. The CRA/CSA instructional sequence is infused throughout each instructional approach included in this book. The rationale for using the CRA/CSA sequence is its long history of successful research and its applicability to address students' processing and language deficits that contribute to their mathematics failure. Finally, this book may be used as a whole guide to interventions across grade-level standards, or teachers may prefer to use specific chapters to enhance their instruction addressing certain standards.

REFERENCES

Baddeley, A. (2000). The episodic buffer: A new component of working memory? *Trends in Cognitive Sciences, 4*, 417–423.

Baddeley, A. (2010). Working memory. *Current Biology, 20*, R136–R140.

Bouck, E. C., & Flanagan, S. M. (2010). Virtual manipulatives: What they are and how teachers can use them. *Intervention in School and Clinic, 12*(3), 168–176.

Bruner, J. S. (1966). Toward a theory of instruction. *Harvard University Press, 36*(3), 337–340.

Harris, C. A., Miller, S. P., & Mercer, C. D. (1995). Teaching initial multiplication skills to students with disabilities in general education classrooms. *Learning Disabilities Research and Practice, 10*(3), 180–195.

Jordan, L., Miller, M. D., & Mercer, C. D. (1998). The effects of concrete to semi-concrete-to-abstract instruction in the acquisition and retention of faction concepts and skills. *Learning Disabilities: A Multidisciplinary Journal, 9*(3), 115–122.

Jordan, N. C., Kaplan, D., Locuniak, M. N., & Ramineni, C. (2007). Predicting first-grade math achievement from developmental number sense trajectories. *Learning Disabilities Research & Practice, 22*, 36–46.

Kolkman, M. E., Kroesbergen, E. H., & Leseman P. P.M. (2013). Early numerical development and the role of non-symbolic and symbolic skills. *Learning and Instruction, 25*, 95–103.

Kroesbergen, E. H., Noordende, J. E. van't, & Kolkman, M. E. (2014). Training working memory in kindergarten children: Effects on working memory and early numeracy. *Child Neuropsychology, 20*, 23–37.

Lesh, R., Landau, M., & Hamilton, E. (1983). Conceptual models and applied mathematical problem-solving research. In R. Lesh & M. Landau (Eds.), *Acquisition of mathematics concepts and processes* (pp. 263–343). New York, NY: Academic Press.

Maccini, P., & Hughes, C, A. (2000). Effects of a problem solving strategy on the introductory algebra performance of secondary students with learning disabilities. *Learning Disabilities Research and Practice, 15*(1), 10–21.

Maccini, P., & Ruhl, K. L. (2000). Effects of a graduated instructional sequence on the algebraic subtraction of integers by secondary students with learning disabilities. *Education and Treatment of Children, 23*(4), 465–489.

Miller, S. P., & Mercer, C. (1992). CSA: Acquiring and retaining math skills. *Intervention in School and Clinic, 28*(2), 105–110.

Miller, S. P., & Mercer, C. (1993). Using data to learning about concrete-semiconcrete-abstract instruction for students with math disabilities. *Learning Disabilities Research and Practice, 8*(2), 89–96.

Miller, S. P., & Morin, V. A. (1998). Teaching multiplication to middle school students with intellectual disabilities. *Education and Treatment of Children, 21*(1), 22–36.

Mononen, R., Aunio, P., & Koponen, T. (2014). A pilot study of the effects of Rightstart instruction on early numeracy skills of children with specific language impairment. *Research in Developmental Disabilities, 35*, 999–1014.

Peng, P., & Fuchs, D. (2016). A meta-analysis of working memory deficits in children with learning difficulties: Is there a difference between verbal and numerical domain? *Journal of Learning Disabilities, 49*, 3–20.

Powell, S. R., Fuchs, L. S., & Fuchs, D. (2013). Reaching the mountaintop: Addressing common core standards in mathematics for students with mathematics difficulties. *Learning Disabilities Research and Practice, 28*, 38–48.

CHAPTER 2

Number Sense and the Concrete-Representational/ Semi-Concrete–Abstract Sequence

OVERVIEW

This chapter shows the importance of number sense (i.e., numeracy), how number sense is developed, and the association of number sense to the implementation of the concrete-representational/semi-concrete–abstract (CRA/CSA) sequence. Some children master number sense concepts through informal learning before entering kindergarten, yet others do not develop an understanding of numbers, the meaning behind numbers, and interrelations among numbers. Researchers show that students who display difficulty with number sense also struggle in learning mathematics concepts later (Locuniak & Jordan, 2008; Toll & Luit, 2014). This chapter will provide readers with an overview of number sense and connect number sense to higher order skill development, as well as show how evidence-based instruction helps shape number sense abilities for students who have difficulty with mathematics.

DESCRIPTION OF NUMBER SENSE

Number sense is defined as "moving from the initial development of basic counting techniques to more sophisticated understandings of the size of numbers, number relationships, patterns, operations, and place value" (National Council of Teachers of Mathematics [NCTM], 2000, p. 79). It is characterized as one's fluidity and flexibility with numbers, the sense of what numbers mean, the ability to mentally solve problems using numbers, estimating, and the ability to make numeric comparisons based on observations

(Gersten, Jordan, & Flojo, 2005). Difficulties with number sense interfere with students' acquisition of mathematical principles and applications (i.e., number operations and fractions) later in school (Mazzocco & Thompson, 2005; Van Luit & Schoman, 2000). Gersten et al. (2012) identified four major elements of numeracy. They are magnitude comparison, strategic counting, the ability to solve simple word problems, and retrieval of basic arithmetic facts. Within mathematics standards, these elements include skills such as counting, number knowledge, number transformation, estimation, and the ability to create and identify number patterns that are also found in the mathematics standards across the United States (Common Core State Standards Initiative, 2010). Younger students in school struggle with counting, cardinality, magnitude, fluency, and basic combinations of numbers (Mazzocco & Thompson 2005; Powell et al., 2013). Each of these skills is tied to understanding numbers. In fact, Kroesbergen, van't Noordende, and Kolkman (2014) explain the most important components of number sense are counting and quantity knowledge.

NUMBER SENSE AND MATHEMATICAL REPRESENTATIONS

Bruner (1966) developed three stages of how people use representations to understand information that develop typically in early childhood. The stages are the enactive stage, iconic stage, and symbolic representation stage. The enactive stage develops in early childhood (infancy to age 6). During the enactive stage, young children manipulate objects but do not develop an internal representation of the objects. Therefore, young children can play with an object but would not have a picture of that object when thinking about it. The iconic stage develops in children starting at age 6. During the iconic stage, children can develop mental images of what they have manipulated. Therefore, children at age 6 can think about objects and visualize them in their mind. The third stage is the symbolic stage, which begins at 7 years of age. During the symbolic stage, information is stored in the form of code or symbols, which can be classified and organized. Therefore, children can create a mental image and organize that image into a category. An example of these stages is shown in Figure 2-1.

Students must have practice with numbers in which they encounter repeated experiences that give them time and opportunity to develop ideas about mathematics, understanding, and increase fluency (Fuson, Clements, & Sarama, 2015). Teachers must provide students with opportunities to build enactive, iconic, and symbolic representations in the facilitation of learning. When developing a sense of what numbers are and what numbers mean, students require experiences in which they manipulate numbers, create mental images of those numbers, and organize the numbers using symbols (Clements, 1999).

Figure 2–1. *Stages of Representation.*

DEVELOPMENT OF NUMBER SENSE AND TRANSLATING MATHEMATIC SYMBOLS

Clements, Sarama, Spitler, Lange, and Wolfe (2011) state that effective instruction for students at risk in mathematics must include knowledge of how to think about a concept paired with instructional tasks designed to move them through levels of thinking in which they make gains toward educational goals. There are times in which students have difficulty making relations with mathematics and organizing symbolic information. Therefore, it is important for teachers to have an understanding of how students translate mathematical symbols to their meaning. According to Dehaene (2001), the cognitive underpinning of number sense comes from three ways of interpreting representations of numbers in one's brain through one's experiences. These interpretations are also referred to as codes because they are the mechanism in which people translate symbolic and nonsymbolic information (Kolkman, Kroesbergen, & Leseman, 2013). These interpretations are made via an analogue quantity code (magnitudes of numbers), a verbal code (the spoken number name), and a visual code (the written number symbol) (Kroesbergen et al., 2014). The analogue quantity code involves a metaphorical or mental number line in which a person can estimate magnitudes of numbers (e.g., more, less). The verbal code involves the number names that correspond to number quantities (e.g., three). The visual code involves the number symbol and written word that represent the quantity (e.g., 3 as written word three). Experiences in which students use analogue, verbal, and visual codes to translate meaning among symbolic and nonsymbolic information build students' awareness of the number system, and this is how connections between the number system and quantities in the environment are made (Berch, 2005). This section explains how students build awareness of numbers that represent quantities along with the influence on mathematical instruction.

MAPPING SKILLS

Number words and number symbols have meaning only if they become associated with the quantities they represent (Geary, 2013). Students' abilities to connect the number symbol to the quantity that symbol represents are called mapping skills. For example, students develop an awareness that the number symbol "5" or "five" matches the amount of five objects, and the number "18" is smaller than "42" (Kolkman et al., 2013). Research literature indicates mapping skills are based on students' ability to translate magnitude of quantities, verbal representations of quantities, and visual representations of quantities of nonsymbolic and symbolic information. These mapping skills are shown in Figure 2–2.

Students from the onset of schooling up to second grade are in the process of mapping symbolic information (i.e., numerical digits and verbal

Nonsymbolic Information

Symbolic Information

Visual Code: Number Symbol	Verbal Code: Number Name	Verbal Code: Counting to Eight
8	eight	One, two, three, four, five, six, seven, eight

Figure 2–2. Mapping Skills.

names of numbers) onto preexisting nonsymbolic representations of whole numbers (Bonny & Lourenco, 2013). Therefore, students are continually making mental linkages and refining existing number awareness between quantities they manipulate to the cultural symbols and numerical expressions that have been assigned to those quantities (Sasanguie, Gobel, Moll, Smets, & Reynvoet, 2013). The ability to understand and manipulate numerical magnitudes is called nonsymbolic skill.

Examples of students demonstrating nonsymbolic skills are identifying quantities as more or less or solving problems using quantities that do not rely on verbal or visual number symbols. When demonstrating nonsymbolic skills, students do not have to connect the number symbol or number name to the corresponding quantity. Symbolic skills are verbal and visual representations. They are culturally based (i.e., numbers words such as thirteen or Arabic number symbols such as 13) acquired skills, which mean that students have to be taught the symbolic expressions. Students demonstrate symbolic skills when they say number words when counting or identify number symbols they see in their environment. When demonstrating symbolic skills, students do not necessarily connect the number symbol to the corresponding quantity (i.e., they do not realize that 5 or five is the same as the quantity of five). In addition, the English language makes it difficult to connect symbolic numbers to the nonsymbolic amounts they represent. For example, in the number 15, the value in the ones place is identified first and "teen" represents the value of 10. This is shown in Figure 2–3.

When young students start school, the process of coding nonsymbolic and symbolic information are separate ways of understanding numbers, which means that students do not necessarily link a number word or number symbol to the quantity that number represents. Therefore, when young students are presented with a number symbol, they interpret it as only symbolic or only nonsymbolic. That is, when students identify the number "9" by

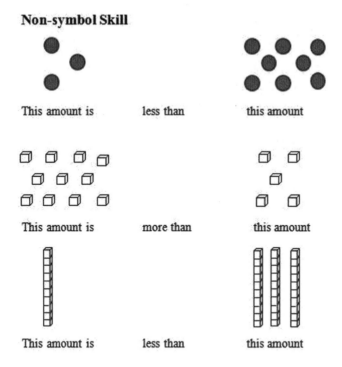

Figure 2–3. Symbolic Numbers and Nonsymbolic Amounts.

saying "nine," they do not necessarily understand that the Arabic number 9 or word "nine" is the same as the quantity of 9. However, through mathematical experiences that allow students to implement mapping skills, translation of nonsymbolic and symbolic information using analog, verbal, and visual codes become more integrated, allowing for automatic processing of numbers. Thus, it is important that the teacher ensures instructional tasks lead to the effective integration of translating symbolic and nonsymbolic information. Moving flexibly between various representations is essential in developing number sense. Making connections among number symbols and quantities is possible through working memory processes, which are discussed in the next section. Demonstrations of mapping symbolic and nonsymbolic information are shown in Figure 2–4.

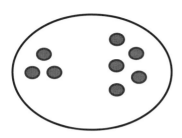
Three and five together make 8 **or** 3 + 5 = 8.
Three is smaller than 8.
Five is smaller than 8.
Eight is the largest number.
Three is the smallest number.
Five is in between three and eight.

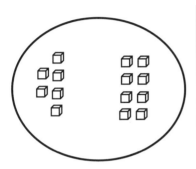
Six and eight are fourteen **or** 6 + 8 = 14.
Six is smaller than eight.
Six is smaller than fourteen.
Fourteen is the largest number.
Six is the smallest number.
Eight is in between six and fourteen.

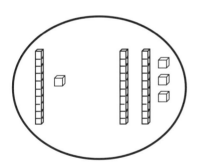
Eleven and twenty-three are thirty-four **or** 11 + 23 = 34.
Eleven is smaller than twenty-three.
Eleven is smaller than thirty-four.
Thirty-four is the largest number.
Eleven is the smallest number.
Twenty-three is in between eleven and thirty-four.

Figure 2–4. *Demonstrations of Mapping Symbolic and Nonsymbolic Information.*

APPLICATION OF CRA/CSA TO TEACH NONSYMBOLIC AND SYMBOLIC SKILLS

The CRA/CSA sequence combines instruction that builds nonsymbolic and symbolic skills in which learning requires students to use visual, auditory, and kinesthetic processes (Witzel, 2005). Students first learn mathematical concepts through the nonsymbolic manipulation of objects and simultaneously apply verbal and visual symbols of numerical representations. Once students have mastered the mathematical concept through the manipulation of objects, instruction moves to a second phase in which students manipulate the mathematical concept through drawings that represent the nonsymbolic quantities and simultaneously apply verbal and visual symbols (i.e., number names or numbers) of the numerical representations. Manipulation of mathematical concepts through the use of drawings acts as a bridge that helps students make linkages from problem-solving mathematical concepts in a combined nonsymbolic and symbolic manner to a symbolic-only manner in which the representation of symbolic numbers is solely used (Flores, 2009). The ultimate goal is for students to move flexibly between all representations, but this sequence of instruction has been found to be an effective scaffolding tool that builds this understanding. An example of nonsymbolic and symbolic problem solving using drawings is shown in Figure 2–5.

Before students solve mathematical concepts using a strictly symbolic method of numbers only, students learn a strategy that helps them coordinate and retrieve steps in the process of thinking about and applying the mathematical concept. Once students learn the strategy, they receive instruction about the mathematical concept with numbers only. Miller (2009) explains CRA/CSA builds conceptual, procedural, and declarative mathematical knowledge. Conceptual knowledge is an understanding of the mathematical concept itself. Procedural knowledge is an understanding of the steps required to use the mathematical concept in problem solving. Declarative knowledge is the ability to solve problems accurately and automatically. The next section describes the phases of CRA/CSA and the implementation of instruction that build conceptual, procedural, and declarative knowledge through simultaneous manipulation of nonsymbolic and symbolic methods.

CRA/CSA APPLICATION AND MORE, LESS, SAME

Instruction for more, less, and the same begins with a short interval (about 5 minutes) of exploration with manipulative counters such as beans, counting bears that are the same size, and the use of a balance scale. The teacher leads a discussion about the scale and vocabulary such as what more means,

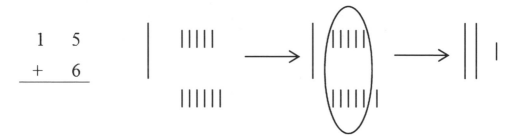

Figure 2–5. Example of Nonsymbolic and Symbolic Problem Solving.

what less means, and what the same means. Before modeling how the scale is used to determine more, less, or the same, the teacher checks for students' understanding of the vocabulary terminology. At this point, student understanding can be as basic as more is larger, less is smaller, and the same means no difference or even.

More, Less, and Same With Objects (Concrete)

The teacher lays out two sets of nonsymbolic amounts for students to see. At first it is helpful if the sets are different colors. For example, one set can be red counters and the other set blue counters. The teacher explains that the counters are in sets and the scale is going to be used to see which set is more. The teacher asks the student to predict which will be heavier and to explain why. Then he places the sets in the scale. When the larger set moves down, the teacher asks the student to explain why the larger item moved down. If the student doesn't state it, it is important for the teacher to explicitly note that the larger set is heavier and will move that side of the scale downward. The larger set is more. The teacher repeats this demonstration for the concepts of less and the same. With each set, students will predict and explain their thinking before the demonstration occurs. After each group is weighed, the teacher will ask the student to explain why the larger set moved down, the smaller set moved up, or they stayed the same. If the student is unable to explain and predict, the teacher can explain the concepts of more, less, and the same. For the concept of less, the teacher can facilitate discussions that the smaller set will move up because it is lighter, and the smaller set is less. To explain the concept of same, the teacher facilitates discussions that neither set will be heavier or lighter; the scale will not move up or down. An example of this use of a balance scale is shown in Figure 2–6.

Once the teacher demonstrates the concepts of more, less, and same, the teacher and students use the scale together. The teacher will lay out amounts and assist the students in identifying more, less, or the same based on what happens to the scale. After several practice sessions with the teacher providing assistance, independent practice is implemented. Students take turns placing amounts in the scale and identifying which amount is more, less, or the same. The teacher provides feedback based on students' responses, prompting students to think about what is happening and why.

More, Less, and Same With Drawings (Representational)

Once students demonstrate understanding of more, less, and same, the teacher incorporates drawing amounts and emphasizes the symbolic number names. For example, the teacher lays out sets, counts the sets using the

20   *Making Mathematics Accessible for Elementary Students Who Struggle*

Think aloud: *We have two groups of counters, one red and one blue. If we put each group on one side of the scale, it will tell us which is more. How will I know if the scale is telling that one group is more? Will that side be lower or higher? It will be lower because more counters will pull it down.*

Think aloud: *Now I have the counters on the scale and I notice what has happened to each side of the scale. If the sides were at the same level, I would have equal amounts. Are they at the same level? No, so the amounts are not equal. If one side is lower than the other, then that group is more. It is pulling that side down. Is one side lower? Then, are there more red counters?*

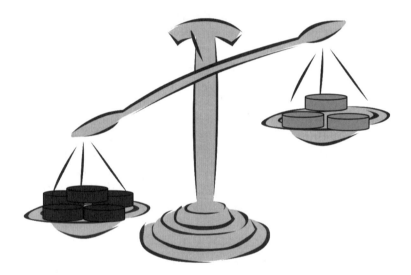

Think aloud: *Now I have described the red counters. What do I notice about the blue counters? Is the side with the blue counters lower? No. So, there are not more blue counters. Is the blue side lower or higher than the side with red counters? It is higher. That means that there are not enough blue counters to be at the same level as the red counters or equal. There are less blue counters than red counters.*

Figure 2–6. *Use of a Balance Scale to Teach Equal.* continues

counting strategies that will be discussed in the next chapter, draws the amounts counted, and then verbally and/or visually assigns the number name and visual symbols of numbers. The teacher facilitates mathematics talk in which communication about which number name and visual number symbol

Think aloud: *We have two groups of counters, one red and one blue. If we put each group on one side of the scale, it will tell us whether the groups are equal, if one is less or more than the other. How will I know if the scale is telling that one group is more? Will that side be lower or higher? It will be lower because more counters will pull it down. What will the scale show if the groups are the equal? Will both sides be the same height from the table? Yes. Let's see.*

Think aloud: *What do I notice about the sides with the red and blue counters? Is the side with the red counters lower? No. So, there are not more red counters. Is the blue side lower than the side with red counters? No, so there are not more blue counters. Are the sides of the scale the same? Yes, so does that tell us the numbers are equal? Yes.*

Figure 2–6. continued

is more, less, and same based on what is placed in the scale or drawn out. After several practice sessions with the teacher providing assistance, students identify more, less, and the same incorporating the symbolic number names and visual symbols on their own. Students take turns placing amounts in the scale and identifying which amount is more, less, or the same. The teacher provides feedback based on students' responses, prompting students to communicate about what is happening using number names and visual symbols. An example of the balance scale used with sets and drawings is shown in Figure 2–7.

Two objects are placed on each side of the balance scale

When the child talks about what he created, he explains that both sides have two and that the amounts are the same because the scale is even and balanced.

Figure 2–7. Use of Balance Scale Used With Sets and Drawings.

More, Less, and Same With Symbols (Abstract)

Once students demonstrate an ability to draw amounts and emphasize the symbolic number names in discussing more, less, and same, the teacher emphasizes primarily number symbols, and drawings or the balance scale are used to check predictions. The teacher facilitates math talk in which communication about which number name and visual number symbol is more, less, and same. The instructor facilitates discussions of why and uses drawings and the scale as a follow-up based on students' responses using numbers only. After several practice sessions with the teacher providing assistance, students identify more, less, and the same using the symbolic number names and visual symbols on their own. Students take turns checking their predictions by placing amounts in the scale or drawing out the amounts. The teacher provides feedback based on students' responses, facilitating communication about what is happening.

CRA/CSA APPLICATION SORTING AND CLASSIFYING NONSYMBOLIC AMOUNTS

Due to the importance of mathematical representations, students must have ample experiences sorting and classifying information and then communicating verbally and visually about the objects, as well as relationships between the objects and mathematical symbolic representations of our number system. This section will describe how to implement sorting and classifying activities within the CRA/CSA sequence. It is important to note that most lessons in sorting and classifying nonsymbolic amounts require students to use counting strategies that will be discussed later in this chapter.

Mathematics Stories and Sorting and Classifying With Objects (Concrete)

Mathematics stories are meaningful experiences teachers facilitate to help students create nonsymbolic amounts and assign meaning to those amounts. Mathematics stories are often paired with read-alouds, songs, and finger plays. It needs to be noted that although good practice, it is not required that the stories are paired to a read-aloud, song, or finger play. When using mathematics stories, the teacher lays a set of a nonsymbolic amount using familiar objects for students to see and helps students create scenarios in which they identify and discuss an amount. For example, the teacher lays out three plastic pigs from the story of the "Three Little Pigs" and explains that each pig had a house. She does not tell students the amount but leads a discussion on how could the class know how many houses in all. In the discussion, the teacher asks students to predict the amount of houses and why. The teacher can even allow students to act out the scenario using the pigs and houses or students can pretend to be the pigs. At the end of the discussion, she models counting the pigs, associating a house (pictures of each house can be used) with each pig, and writes out the number in symbolic form. The teacher and students then create another story together, and students are given the opportunity to communicate the amounts using number words and number symbols. The teacher scaffolds students' understandings about the amounts by asking questions, helping students count, and assisting students if they have difficulty communicating or writing the symbolic forms of the amounts. After the teacher and students create a story together, students create a story on their own.

The stories slowly become more complicated in which students sort objects through attributes such as colors, size, and shape through the use of stories and communicate amounts of objects that were sorted using symbolic number forms. With enough experience, the teacher can fold in many sets of objects that have different attributes and have students communicate how they organized the sets and assigned a symbolic quantity to each set. For example, using the song "Five Little Monkeys Jumping on the Bed," the teacher can stop in the middle of the song when three monkeys are on the bed and two monkeys are off the bed, then facilitate a discussion of how many monkeys are on the bed and how many are off. The teacher allows students to make predictions of how many and discuss how they would put the monkeys in different groups. After the discussion, the teacher uses finger puppets and acts out sorting the monkeys on and off the bed. The teacher then models verbally counting the monkeys in each group and writes out the number in symbolic form. The teacher and students together create another scenario in which the monkeys are put into groups using the song. Students are given the opportunity to communicate the amounts to describe the groups using number words and number symbols. The teacher continually scaffolds students' understandings through asking questions, helping students count, and assisting students if they have difficulty communicating or writing the symbolic forms of the amounts. After the teacher and students create scenarios with the song together, the teacher invites students to create a story in which students put monkeys into different groups and communicate about the groups and amounts using symbolic forms of numbers.

Mathematics Stories Using Drawings (Representational)

Once students have ample practice creating mathematics stories, they can communicate their thoughts about amounts through drawings in journals. When students are beginners in journaling, the teacher still provides an amount using concrete objects. Going back to the example of the three pigs and houses, the teacher lays out the three pigs and leads a discussion on how would we know how many houses in all. She allows students to make predictions. After the discussion, the teacher models associating a house (pictures of each house can be used) with each pig and draws out pictures of the pigs and houses in her journal. Then she models counting and writes out the number in symbolic form. The teacher and students then need to create another story together, and students communicate the amounts verbally and writing in their journals using pictures along with number words and number symbols. The teacher scaffolds students' understandings about the amounts by asking questions, helping students count, and assisting students if they have difficulty communicating or writing the symbolic forms of the numbers. After the teacher and students create a story together, students create a story independently using journals to communicate their ideas in addition to verbally describing or identifying numbers.

Journals become more complicated when students are asked to draw out how they grouped the concrete representations of objects based on attributes and different groupings of sets. Teachers can push thought further by having students draw out and describe which group has more, less, or same, and why. It is important that teachers still provide students with concrete objects that they sort and organize in addition to having students create drawings, then discuss and write the amounts using symbolic forms of numbers. It is also important for teachers to model creating and drawing different groupings of objects, create and draw different groupings of objects with students, and allow students to create and draw different groupings of objects independently.

Mathematics Stories Using Symbols (Abstract)

Once students have ample practice drawing out mathematics stories, they can communicate their thoughts about amounts with an emphasis on the number symbols themselves. Once again using the example of the three pigs and houses, the teacher lays leads a discussion on the three little pigs and houses then writes the number symbols. She allows students to make predictions. After the discussion, the teacher verbally models associating a house with each pig. Then she models checking her prediction by drawing out pictures of the pigs and houses. The teacher and students then create another story together, and students communicate the amounts verbally using the symbols and check their predictions using pictures along with number words and number symbols. The teacher scaffolds students' **understanding** through guided questions, helping students count, and assisting students if they have difficulty communicating or writing the symbolic forms of the numbers. After the teacher and students create a story together, students create a story independently using journals to communicate their ideas using numbers only and check their predictions through drawings or objects if students need more assistance.

CHAPTER SUMMARY

Students who have difficulty with number sense also struggle in learning mathematics concepts later (Locuniak & Jordan, 2008; Toll & Luit, 2014). This chapter described the development of fluidity and flexibility with numbers and understanding that leads to skills related to mental computation, estimation, and numeric comparison, which are critical outcomes of number sense (Gersten et al., 2005). Language development and memory are intertwined with the development of number sense, and students who struggle in mathematics may also have deficits associated with these processing skills. Therefore, interventions should provide explicit instruction related to language and thinking through modeling and guidance in thinking aloud and verbal description of number representations. Although the development of number sense begins in early grades, it extends to more complex concepts and forms the basis for all other mathematics learning. For example, understanding of the concept of equal extends through algebra and beyond. It is crucial that teachers provide opportunities for students to discuss and show their understanding of number concepts rather than assuming mastery based on rote tasks such as recitation of numbers or naming quantities. Teachers may discover poor number sense in students in intermediate elementary grades, and it is never too late to address them. For example, students who struggle with regrouping often do so based on a poor sense of numbers. Providing students with concrete and representational/semi-concrete experiences when teaching these operations using explicit instruction is one solution that research has shown to be effective.

APPLICATION QUESTIONS

1. Describe how memory and language processing are related to the development of number sense.

2. Within the context of teaching students what is represented by the word "eight," describe each of the following: concrete instruction, representational/semi-concrete instruction, and abstract instruction.

3. A lesson includes the composition of eight and its relation to three and five at the concrete level using objects. How would you implement each of the explicit instructional steps to teach this concept (advance organizer, model, guided practice, independent practice, postorganizer)?

4. What is the rationale for balance scales for teaching the concepts of more, less, and same? How does this tool benefit students?

5. Provide reasons why and how a teacher would use a mathematics journal.

6. How would a teacher think aloud during a mathematics story at the concrete level, representational level, and abstract level?

REFERENCES

Berch, D. B. (2005). Making sense of number sense: Implications for children with mathematical disabilities. *Journal of Learning Disabilities, 38*, 333–339.

Bonny, J. W., & Lourenco, S. F. (2013). The approximate number system and its relation to early math achievement: Evidence from the preschool years. *Journal of Experimental Child Psychology, 114*, 375–388.

Bruner, J. S. (1966). *Toward a theory of instruction* (Vol. 59). Cambridge, MA: Harvard University Press.

Clements, D. H. (1999). Subitizing: What is it? Why teach it? *Teaching Children Mathematics, 5*, 400–405.

Clements, D. H., Sarama, J., Spitler, M. E., Lange, A. A., & Wolfe, C. B. (2011). Mathematics learned by young children in an intervention based on learning trajectories: A large scale cluster randomized trial. *Journal of Research in Mathematics Education, 42,* 127–166.

Common Core State Standards Initiative (CCSSI). (2010). *Common Core State Standards for Mathematics.* Washington, DC: National Governors Association Center for Best Practices and the Council of Chief State School Officers. Retrieved from http://www.corestandards.org/assets/CCSSI_Math%20Standards.pdf

Dehaene, S. (2001). Precis of the number sense. *Mind and Language, 16,* 16–36.

Flores, M. M. (2009). Teaching subtraction with regrouping to students experiencing difficulty in mathematics. *Preventing School Failure, 53*(3), 145–152.

Fuson, K. C., Clements, D. H., & Sarama, J. (2015). Making early math education work for all children. *Phi Delta Kappan, 97,* 63–68.

Geary, D. C. (2013). Early foundations for mathematics learning and their relations to learning disabilities. *Current Directions in Psychological Science, 22,* 23–27.

Gersten, R. Clarke, B., Jordan, N. C., Newman-Gonchar, R., Haymond, K., & Wilkins, C. (2012). Universal screening in mathematics for the primary grades: Beginnings of a research base. *Exceptional Children, 78,* 423–445.

Gersten, R., Jordan, N. C., & Flojo, J. R. (2005). Early identification and interventions for students with mathematics difficulties. *Journal of Learning Disabilities, 38,* 293–304.

Kolkman, M. E., Kroesbergen, E. H., & Leseman P. P.M. (2013). Early numerical development and the role of non-symbolic and symbolic skills. *Learning and Instruction, 25,* 95–103.

Kroesbergen, E. H., Noordende, J. E. van't, & Kolkman, M. E. (2014). Training working memory in kindergarten children: Effects on working memory and early numeracy. *Child Neuropsychology, 20,* 23–37.

Kroesbergen, E. H., & Van Luit J. E. H. (2003). Mathematics interventions for children with special educational needs a meta-analysis. *Remedial and Special Education, 24,* 97–114.

Locuniak, M. N., & Jordan, N. C. (2008). Using kindergarten number sense to predict calculation fluency in second grade. *Journal of Learning Disabilities, 41,* 451–459.

Mazzocco, M. M., & Thomspon, R. E. (2005). Kindergarten predictors of math learning disability. *Learning Disabilities Research & Practice, 20,* 142–155.

Miller, S. P. (2009). *Validated practices for teaching students with diverse needs and abilities* (2nd ed.). Upper Saddle River, NJ: Pearson.

National Council of Teachers of Mathematics. (2000). *Principles and standards for school mathematics.* Reston, VA: Author.

Powell, S. R., Fuchs, L. S., & Fuchs, D. (2013). Reaching the mountaintop: Addressing common core standards in mathematics for students with mathematics difficulties. *Learning Disabilities Research and Practice, 28,* 38–48.

Sasanguie, D., Gobel, S. M., Moll, K., Smets, K., & Reynvoet, B. (2013). Approximate number sense, symbolic number processing, or number-space mappings: What underlies mathematics achievement? *Journal of Experimental Child Psychology, 114,* 418–431.

Toll, S. W. H., & Van Luit, J. E. H. (2014). Effects of remedial numeracy instruction throughout kindergarten starting at different ages: Evidence from a large scale longitudinal study. *Learning and Instruction, 33,* 39–49.

Van Luit, J. E. H., & Schopman, E. A. M. (2000). Improving early numeracy of young children with special educational needs. *Remedial and Special Education, 21,* 27–40.

Witzel, B. S. (2005). Using CRA to teach algebra to students with math difficulties in inclusive settings. *Learning Disabilities: A Contemporary Journal, 3,* 49–60.

CHAPTER 3

Counting, Cardinality, and the Concrete-Representational/ Semi-Concrete–Abstract Sequence

OVERVIEW

This chapter shows how the concrete-representational/semi-concrete–abstract (CRA/CSA) sequence is used to facilitate learning of counting and cardinality for young students. Some students master counting skills through informal learning before entering school, but others do not develop an understanding of counting without instruction. Researchers show that students who display difficulty with numbers display smaller learning gains in mathematics throughout schooling and struggle in mathematics concepts later (Jordan, Kaplan, Locuniak, & Ramineni, 2007). This chapter provides readers with an overview of counting principles, the developmental sequence of counting, an overview of developmentally appropriate practice and CRA/CSA, and implementation of CRA/CSA to teach young students with mathematic difficulties how to count and build understandings of quantities.

COUNTING PRINCIPLES

One way young students build number sense awareness is through counting. Counting is the most basic and important skill of number sense (Kroesbergen, van't Noordende, & Kolkman, 2014). Gelman and Gallistel (1978) describe five fundamental principles students must have to be successful in counting. All five principles embody the notion of conservation of numbers. Conservation of numbers means that children conserve numbers when they realize that sets of numbers remain equivalent regardless of their arrangement (Piaget & Szeminska, 1952).

The first principle is one-to-one correspondence. One-to-one correspondence asserts that one and only one number is assigned to each object in a set. The second principle is the stable order principle. The stable order principle asserts that number words always progress in the same order. The third principle is the cardinal principle. The cardinal principle asserts that the last number word counted represents the sum of the set. The fourth principle is the irrelevance principle. The irrelevance principle asserts that object order for counting does not influence the final number counted for the set. The fifth and final principle is the abstraction principle. The abstraction principle asserts that the principles of one-to-one correspondence, stable order, cardinality, and irrelevance apply when counting any set, regardless of what the collection of objects look like.

The most important components of number sense are counting and quantity knowledge (Kroesbergen et al., 2014). Students are expected to demonstrate counting and cardinality at the end of their kindergarten year (CCSSI, 2010). According to most standards, young students demonstrate counting skills by counting to 100 using ones and tens, by applying the principles of counting, and by comparing numbers. Students in kindergarten who display lower mathematics performance struggle with counting, cardinality, magnitude, fluency, and basic combinations of numbers and make smaller gains in mathematics throughout their schooling (Jordan et al., 2007; Mazzocco & Thompson 2005; Powell et al., 2013). Instructional tasks must integrate symbolic and nonsymbolic information and assist students in the development of counting. Effective instruction for young students creates experiences in which students apply verbal and visual number symbols to quantities (Kolkman et al., 2013).

SEQUENTIAL DEVELOPMENT OF COUNTING

The most common numerical activity in preschool is counting (Ramani & Siegler 2011). There are stages of development that outline the learning trajectory of students' counting skills (Clements & Sarama, 2010). Van de Rijt and Van Luit (1998) explain the developmental process of learning to count and the order in which counting skills develop. These developmental stages reflect the trajectories of learning outlined by Clements and Sarama. Each stage of counting conveys how students quantify amounts and make connections with number symbols and number words. The stages are as follows:

1. Acoustic Counting, also referred to as Chanter
2. Asynchronous Counting, also referred to as Reciter
3. Synchronous Counting, also referred to as Corresponder
4. Resultative Counting, also referred to as Counter or Producer
5. Shortened Counting

The stages of counting are described along with an illustration. Links among the stages and principles of counting are also described. Flexible counting is an essential foundation in mathematics and is the goal of counting instruction. Students first learn acoustic counting around the age of 3. Acoustic counting involves speaking numbers but not connecting numbers with quantities of objects. Students who demonstrate acoustic counting are also referred to as chanters because they typically demonstrate acoustic counting through simple songs or rhymes. Examples of acoustic counting are shown in Figure 3–1.

Children's Rhymes That Foster Counting

One, Two Buckle My Shoe

One, two, buckle my shoe. Three, four, shut the door. Five, six, pick up sticks. Seven eight, shut the gate. Nine, ten, begin again.

This Old Man

This old man, he played one. He played knick-knack on my thumb. Knick-knack paddy whack, give the dog the bone. This old man came rolling home.

This old man, he played two. He played knick-knack on my shoe. Knick-knack paddy whack, give the dog the bone. This old man came rolling home.

This old man, he played three. He played knick-knack on my knee. Knick-knack paddy whack, give the dog the bone. This old man came rolling home.

This old man, he played four. He played knick-knack on my door. Knick-knack paddy whack, give the dog the bone. This old man came rolling home.

This old man, he played five. He played knick-knack on my hive. Knick-knack paddy whack, give the dog the bone. This old man came rolling home.

This old man, he played six. He played knick-knack with some sticks. Knick-knack paddy whack, give the dog the bone. This old man came rolling home.

This old man, he played seven. He played knick-knack up in Heaven. Knick-knack paddy whack, give the dog the bone. This old man came rolling home.

This old man, he played eight. He played knick-knack on my gate. Knick-knack paddy whack, give the dog the bone. This old man came rolling home.

This old man, he played nine. He played knick-knack on my spine. Knick-knack paddy whack, give the dog the bone. This old man came rolling home.

This old man, he played ten. He played knick-knack once again. Knick-knack paddy whack, give the dog the bone. This old man came rolling home.

Figure 3–1. *Examples of Acoustic Counting & Children's Rhymes That Foster Counting.*

After acoustic counting, students count asynchronously. Counting asynchronously is when students realize that numbers are used to count quantities of objects. However, they are not able to point to one object while enumerating one number. Students who demonstrate asynchronous counting are also referred to as reciters because they repeat number words but can miss an object or point to the same object twice while counting. An example of asynchronous counting is shown in Figure 3–2.

The next stage is counting synchronously. Students who demonstrate synchronous counting are also referred to as corresponders because students count and enumerate the number at same time. This shows that the student is making a one-to-one relation that is application of the principle of one-to-one correspondence. Students count synchronously around the age of 4 or 5. Students may demonstrate one-to-one correspondence in this stage but can still identify an incorrect amount. An example of synchronous counting is shown in Figure 3–3.

After synchronous counting, students demonstrate resultative counting, or seriation. In resultative counting, students demonstrate the principle of stable order and cardinality. Students who demonstrate resultative counting are also referred to as counters or producers because they accurately count

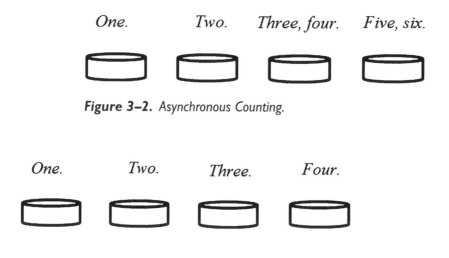

Figure 3–2. Asynchronous Counting.

Think aloud: *I am going to count. Each time I touch a counter, I will say a number. There is one touch per number.*

Figure 3–3. Synchronous Counting.

objects or pictures and can answer how many. Students who understand that counting is relevant to circumstances in which a certain number must be created are producers. Resultative counting means students begin counting with the number one, every object is counted once, and the last number name when enumerating the number is the total number of objects. An example of resultative counting is shown in Figure 3–4.

Finally, students learn shortened counting. Shortened counting requires students to apply the principle of irrelevance. In application of the principle of irrelevance, children must recognize the representation of a number. Therefore, in shortened counting, students count on from a representation of a number they see. For example, in a pair of dice, the student would see two dots on one piece and three dots on the other piece. Instead of touching each dot, the student would say two and continue to count the remaining dots on the other piece. Once students become fluent in shortened counting, they count in a flexible ways. An example of shortened counting is shown in Figure 3–5.

CRA/CSA APPLICATION TO COUNTING INSTRUCTION

Developmentally appropriate practice (DAP) is an approach to teaching that is based on how children learn and develop and what researchers show as effective early education that promotes optimal learning and development (National Association for the Education of Young Children [NAEYC], 2009a). There are three core considerations of DAP and 10 suggested DAP teaching strategies for educators while young students are engaged in learning processes (NAEYC, 2009b). The CRA/CSA sequence is amiable to both in the

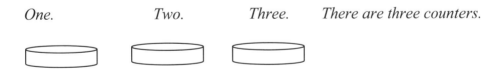

Think aloud: *I am going to count. Each time I touch a counter, I will say a number. There is one touch per number. The last number I say is the total number of counters.*

Figure 3–4. Resultative Counting.

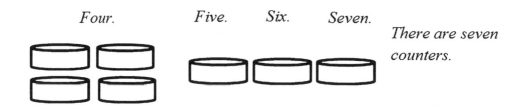

Think aloud: *I am going to count the short way. I know that I have three counters and some more. I start with three and then touch and count starting with the next number after three. Each time I touch a counter, I will say a number. There is one touch per number. The last number I say is the total number of counters.*

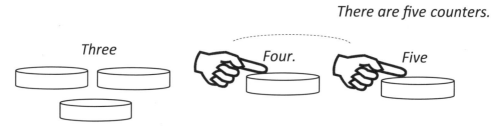

Figure 3–5. Shortened Counting.

implementation of mathematics instruction, making it an effective tool to address the needs of young students with mathematical difficulties.

The DAP core considerations help teachers create learning experiences for children. The first consideration is using knowledge of child development to make decisions on learning experiences that promote learning and development. Teachers can adapt CRA/CSA to assist students with mathematical difficulties through the stages of counting based on the development of counting previously outlined by researchers (Clements & Sarama, 2010; Van de Rijt & Van Luit, 1998). The sequence also allows for integration of nonsymbolic and symbolic skills recommended by researchers (Kolkman et al., 2013). When implementing CRA/CSA, teachers and students use nonsymbolic representations of quantities with objects and then drawings while simultaneously counting and enumerating amounts using symbolic representations of the number names (e.g., five) or number symbols (e.g., 5).

The second consideration is to appreciate individual differences of each child and using that knowledge to build on his or her interests, abilities, and development. When implementing the CRA/CSA sequence, teachers can choose instructional materials for each phase of instruction (i.e., concrete, representation, and abstract) that reflect each student's interests, abilities, and development. It is even possible for students to choose which materials they would like to use during mathematics instruction. As mentioned previously, teachers can also tailor students' learning experiences based on each student's abilities. The third consideration is to be mindful of each child's family values and creating learning experiences that are meaningful, respectful, and relevant. The CRA/CSA sequence is an ideal way for students to build relevant and meaningful experiences in mathematics because it allows students to literally manipulate and problem solve quantities found in their environment. It is important for teachers to respect families and ensure that all materials used in CRA/CSA instruction value the beliefs of the families. For example, certain edible items may not be appropriate. Also, be wary of items with particular caricatures that may represent a television show or movie that a family may not find appropriate. On the other hand, a student may be particularly engaged when counting items that reflect a particular interest or the family's hobby.

The DAP suggested strategies are to be implemented in tandem with the core considerations. The teacher is to choose the strategy that fits a particular instructional situation. The 10 strategies are as follows.

1. Acknowledge
2. Encourage
3. Provide specific feedback
4. Model
5. Demonstrate
6. Create or add challenge
7. Ask questions
8. Give assistance
9. Provide information
10. Give directions

The CRA/CSA sequence is typically coupled with explicit instruction (Miller, 2009). Explicit instruction is recommended by researchers in the provision of instruction for young students with disabilities or who have mathematical difficulties. Explicit instruction implemented in CRA/CSA also encompasses the 10 DAP strategies. The steps of explicit instruction include (a) provide an advance organizer, (b) demonstrate and model the skill, (c) provide guided practice, (d) provide independent practice, and (e) provide a postorganizer (Miller, 2009; Peterson, Mercer, & O'Shea, 1988).

Evidence-based practices for young students involve facilitating play, engagement, and instruction that occur in natural environments using developmentally appropriate practices and universal design (Mogharreban & Bruns, 2009). Specifically in preschool settings for students who are at risk or who have disabilities, teachers need to implement unstructured and structured play coupled with explicit instruction as it applies to young learners (Stanton-Chapman, Denning, & Jamison, 2012). CRA/CSA provides a way of structuring supplemental mathematics instruction for young students who need additional help.

CRA/CSA APPLICATION FOR COUNTING AND CARDINALITY

Counting lessons consist of using concrete objects, and then lessons with pictures of objects are implemented. Lessons build on each other and became more complex by including larger quantities. It is important to start with small numbers such as counting quantities between 1 and 5. Then move to zero, then introduce amounts of 6 and 10. After students have mastered counting numbers to 10, the instructor can move on to counting quantities between 11 and 20. When administering lessons counting quantities between 11 and 20, the teacher needs to have knowledge of place value instruction, which is discussed in the following chapter. Once students demonstrate one-to-one correspondence and cardinality, the teacher provides instruction on counting using shortened counting. After students learn how to count using shortened counting, skip counting can be introduced. Descriptions of how to implement lessons are provided in the next section.

Before teaching students how to count with the expectation of teaching one-to-one correspondence and cardinality, teachers review counting orally to 10. After reviewing and ensuring students can count to 10, the teacher facilitates a discussion that counting is a way of knowing how many. To set clear expectations and limit confusion for beginners, the teacher explains that when they count for this particular lesson, they start with the number 1 and the last number they say is the amount they have.

Concrete Phase One to One and Cardinality

The teacher then lays out an amount of objects (between two and five at first) and has a mat. The objects are off the mat. The teacher models counting by placing the objects on the work mat one by one and stating the respective number as the object was placed on the mat. It is important to place the objects on the mat and enumerate the number because it is a visual for students to connect one-to-one correspondence to the act of counting. Another way to teach this at the concrete level is to set up the mat in a way that is similar to a ten frame, a model used within typical general education instruction. The cells within a frame can be confusing to students as evidenced by students' inclusion of the frame cells within their counting. By beginning instruction at the concrete level, using this model may decrease confusion later. A diagram of this process in shown in Figure 3–6.

After modeling, the teacher clears the mat, lays out another amount of objects on the mat and encourages students to count with him. When the teacher and students count together, the teacher scaffolds understanding by pausing and allowing students to verbally enumerate the symbolic number name as the objects are placed on the mat. With ample practice in which the teacher scaffolds students' understanding saying the number name while objects are placed on the mat and pausing to give students a

Mat and Counters & Mat with Ten Frame and Counters

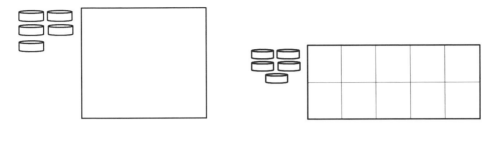

Begin Counting

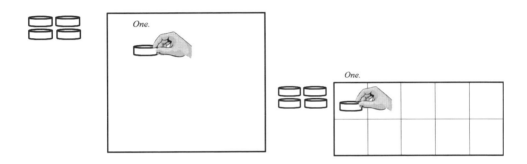

Finish Counting

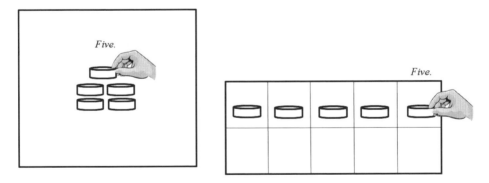

Figure 3–6. Physical Process in Teaching One to One and Cardinality.

chance to say the name, the teacher invites students to count amounts on their own. The teacher gives students objects and mats in which students place objects on the mats and say the number name. The teacher provides feedback based on students' responses, prompting students to communicate using number names. The teacher would also ask students to demonstrate how they counted the objects.

Representational/Semi-Concrete Phase One to One and Cardinality

After counting using objects, the teacher moves counting lessons to the representational/semi-concrete phase in which the lessons are similar, but students count pictures instead of objects. This phase involves counting pictures of objects and counting using ten frames. It is important that the teacher still models counting; invites students to count with him, providing students guidance so that students are successful; and then allows students to count on their own with feedback and questions. A diagram is shown in Figure 3–7.

Ten Frame With Pictures (counting three and zero objects)

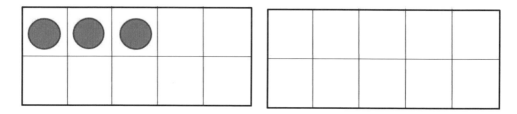

Begin Counting

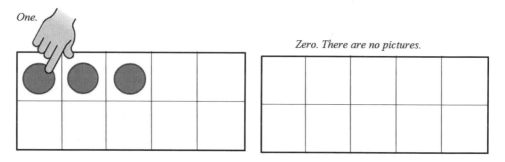

Finish Counting

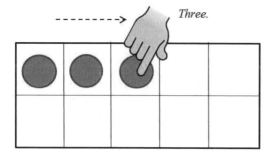

Figure 3–7. Process for One to One and Cardinality at Representational Level.

The teacher can introduce students to counting using ten frames when students successfully count amounts one to five using pictures and demonstrate an understanding of zero. At first, students must have practice placing objects in the ten frame and enumerating the number, as shown within concrete instruction. If they do not have practice placing objects in the ten frame, they may not see the open spaces as zero. Therefore, students will become confused. Once students demonstrate the ability to count amounts placing objects in the ten frame independently, the instructor can move to the representational/semi-concrete phase of using ten frames to count, as shown in Figure 3–7.

Abstract Phase One to One and Cardinality

When students can count amounts using the representational/semi-concrete phase of ten frames, students can write, trace, or identify the symbolic name or number of the amounts they counted. Once students successfully enumerate the amounts while counting demonstrating one-to-one correspondence, the teacher asks students to answer questions such as how many. When students demonstrate one to one counting the amounts, the teacher can model counting and then model writing the number name along with the number symbol.

After modeling, the instructor provides opportunities for the students and teacher to count and write together. Some students will not be able to write the number name or number symbol. Instead of requiring them to write, students can trace or identify the number names or number symbols after counting. The teacher may have to scaffold instruction by modeling counting again for the students and emphasizing the last number stated. If the students cannot answer, then the teacher says the amount and gives the students an opportunity to count again as a way to ensure success. Following several sessions of counting and writing with teacher assistance, students can count an amount using objects and write, trace, or identify the number word and/or symbol independently. The teacher will provide students feedback based on their responses and ask questions about how they counted and how many. Instruction may have to be scaffold, in which the teacher remodels counting and asks guiding questions.

CRA/CSA APPLICATION FOR SHORTENED COUNTING

Before teaching students to count using shortened counting, graphic counting has to be introduced. Graphic counting is referred to as subitizing. It is how an individual identifies amounts without touching the objects or pictures (Clements, 1999). Graphic counting provides another way for students to determine how many and prepare students for seeing numbers in flexible ways. For example, the amount of four could be two and two or one and three. When introducing students to graphic counting, the quantities between one and five should be used. It is recommended to use the amount of three.

Concrete Phase Graphic Counting

The teacher begins the lesson series using objects and a work mat. The teacher says she knows there are three items on the mat and facilitates a discussion that there are ways of counting without having to touch each item. After the discussion, the teacher touches two of the items and says two, and then she touches the last item and says one. She asks students, "If I have two and one, how many do I have?" Students provide responses, and then the teacher says, "Let's find out." She touches the items and counts them showing that there are three. Then she asks students if they can find the amount of two and the amount of one within three. If students do not respond, she models by touching two items, saying, "Here are two," and then touches one item, saying "Here is one. Whenever I see two and one, I know I have three."

After modeling, the teacher would clear the mat, and then place three items on the mat again (the same amount used to model) and say, "Let's find out together how many objects there are without having to touch each item." The teacher gives time for students to explore ways of finding out how many without having to touch each object. During the practice session, instruction includes scaffolding in which the instructor models and asks questions to help students build understanding. Noticing patterns helps students notice the amount. Assisting students with remembering these patterns might involve a metaphor such as *taking a picture in your head*. Ask students to look at the objects that have been identified as a particular number amount, take a picture in their head, and then close their eyes and *look* at the picture in their mind. After completing counting sessions together, the teacher gives students mats and objects and invites students to count objects independently. The teacher consistently checks for understanding and provides feedback to build students' knowledge base in counting. To review, the teacher asks students to demonstrate how they count without having to touch each item. An example of graphic counting is shown in Figure 3–8.

Representational Phase Graphic Counting

Once students are given ample practice counting amounts with objects, instruction includes counting pictures instead of objects. It is important that the teacher still models counting without having to touch each picture; invites students to count, providing them guidance so that students are successful; and then allows students to count without touching each picture on their own with feedback and questions that build understanding. An example of graphic counting with flashcards at the representational level is shown in Figure 3–9.

The representational phase also involves counting pictures within ten frames. Students need to count without touching each picture using ten frames

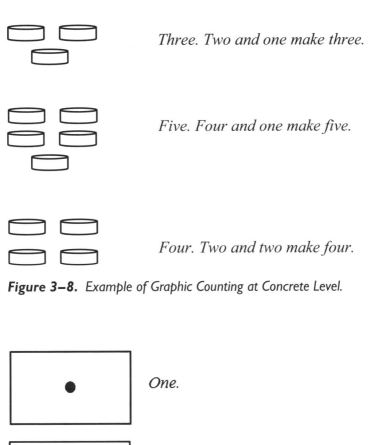

Figure 3–8. Example of Graphic Counting at Concrete Level.

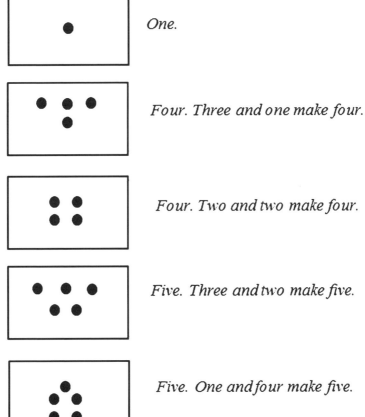

Figure 3–9. Flashcards With Dots for Graphic Counting.

when they successfully count amounts one to five using pictures and demonstrate an understanding of zero. The use of ten frames may allow students to better identify patterns in the pictures. An example of graphic counting with ten frames in shown in Figure 3–10.

Abstract Phase Graphic Counting

When students say accurate amounts to five without having to touch each picture of an object, students can write, trace, or identify the symbolic name or number of the amounts they counted. The teacher can model identifying

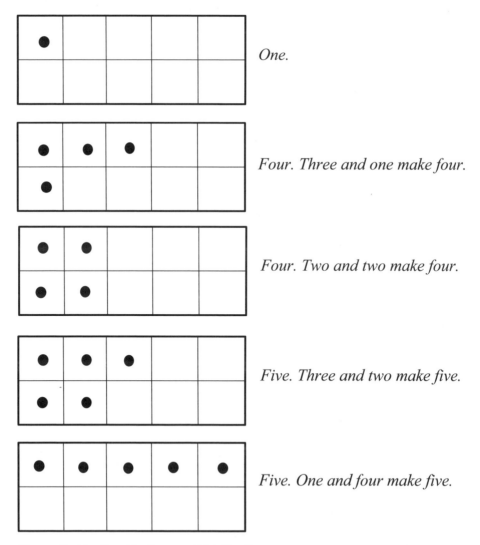

Figure 3–10. Graphic Counting With Ten Frame.

amounts without touching each picture and then model writing the number name along with the number symbol. After modeling, the instructor provides opportunities for the students and teacher to count and write together. Once again, some students will not be able to write the number name or number symbol. Instead of requiring them to write, students can trace or identify the number names or number symbols after counting. The teacher may have to scaffold instruction by modeling counting without touching pictures again for the student and emphasizing the last number stated. If the student cannot answer after the teacher models counting without touching each picture, the teacher should say the amount and then give the student an opportunity to count without touching again. After several sessions of counting without touching the pictures and writing with teacher assistance, students can count an amount using pictures and write, trace, or identify the number word and/or symbol independently. The teacher will provide students feedback based on their responses and ask questions about how they counted and how many. The teacher can scaffold instruction by remodeling counting without touching each picture and asking guiding questions.

Concrete Phase Shortened Counting

Instruction for shortened counting combines graphic counting with resultative counting. Students are given objects in a bowl or mat and are also given objects that go outside the bowl or mat. The teacher facilitates a discussion about counting objects in the bowl without touching, then touching and counting objects outside the bowl. An example of shortened counting at the concrete phase in shown in Figure 3–11.

The teacher can model shortened counting by saying, "I see two and one so I know there are three in this bowl. I will now touch and count the

Think Aloud:

Two and one make three. In the bowl, there are three.

Next, I touch and count four, five.

Figure 3–11. Example of Concrete Phase Shortened Counting.

rest to see how many in all." After modeling, the teacher and students count together. During practice sessions in which the teacher and students count together, the teacher models counting items in the bowl without touching and then continues enumerating the number as he touches and counts each item outside the bowl. The teacher scaffolds instruction through prompting and asking questions that help students build understanding. After counting together, the teacher gives students bowls and objects and invites students to count independently. The teacher pushes thought by requesting students to demonstrate how they count using shortened counting.

Representational/Semi-Concrete Phase Shortened Counting

After practice implementing counting objects in the bowl without touching, then touching and counting objects outside the bowl, instruction moves to using pictures. The teacher still models the graphic and resultative counting and invites students to count with her, providing them guidance so that students are successful, and then allows students to count using combined graphic and resultative counting on their own with feedback and questions that build understanding. Examples of shortened counting in the representational/semi-concrete phase are shown using simple drawings, and ten frames are shown in Figure 3–12.

Abstract Phase Shortened Counting

When students say accurate amounts without having to touch each picture of an object and simultaneously demonstrate resultative counting with the remainder of items, students can write, trace, or identify the symbolic name or number of the amounts they counted. The teacher models counting and writing the number name along with the number symbol. Some students cannot write the number name or number symbol. Instead of forcing students to write, they can trace or identify the number names or number symbols after counting.

CRA/CSA APPLICATION AND SKIP COUNTING

Concrete Phase Skip Counting

Once students demonstrate shortened counting without the need of teacher assistance, skip counting can be introduced. Skip counting can involve any amount but at minimum includes counting in sets of 2, 5, 10, and 3. Beginning lessons start with manipulatives and plates or ten frames. Plates or ten

Think Aloud:

Two and two make four. In the box, there are four.

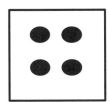

Next, I touch and count five, six, seven.

Two and two make four. In the box, there are four.

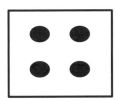

Next, I touch and count five, six, seven.

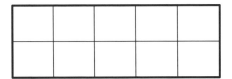

Figure 3–12. *Examples of Shortened Counting Within Representational/Semi-Concrete Phase.*

frames are encouraged more than bowls so students can see every object accounted for in the larger quantities (e.g., 10) they will skip count. The teacher lays out plates that represent sets and places an equal amount of objects on each plate. Examples showing how to set up these concrete experiences are in Figure 3–13.

Plates and Objects

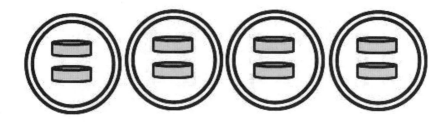

Tens Frames

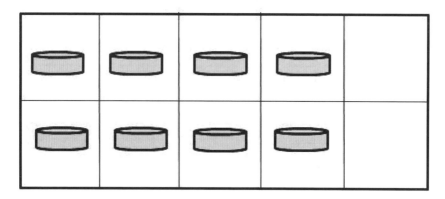

Figure 3–13. Concrete Models for Skip Counting.

The teacher facilitates a discussion about sets and equal numbers being the same amount for each set. She demonstrates what that means by laying out equal amounts and sets for students to see. She then explains that she can determine how many in all by counting each set and that she can count without having to touch each item. She then models counting the sets. After counting, the teacher can write an equation showing that represents the repeated addition she just modeled. If students are not ready to connect an equation to skip counting, the teacher does not have to implement that part of the instruction. Once she models counting the sets and optionally writing an equation, the teacher invites students to count with her. If she chooses for students to write equations, she provides assistance for students who need help. After ample practice counting together, the teacher gives students plates or ten frames and objects, and invites students to count independently. The teacher pushes thought by requesting students to demonstrate how they counted sets to determine how many in all and encourages students to write out equations that show what they did.

Representational/Semi-Concrete Phase Skip Counting

After practice skip counting using objects, instruction moves to using pictures. Pictures include a number line or ten frames that are drawn out. A discussion is facilitated about equal sets and counting items in each set to determine how many in all. The teacher shows students the number line and lays out objects with the number line so students can make the connection that the number on the line represents the amount that will be counted. After practice using the number line with objects, the objects can be removed. If the teacher uses ten frames, the circles or pictures should be in each frame showing an amount. The teacher models skip counting so students can see an example. Modeling and requesting students to write equations is optional. After modeling, the teacher invites students to count with her, providing them guidance so that students are successful. After several practice sessions in which the teacher provides guidance, she prompts students to count sets on their own with feedback and questions that build understanding. She can invite students to write equations to increase the difficulty level. Representational/semi-concrete level models using the number line and a ten frame are shown in Figure 3–14.

Abstract Phase Skip Counting

Once students show an understanding of skip counting using pictures, instruction moves to using numbers only. Instruction of skip counting with numbers only involves using a hundreds chart, a number line with numbers only, or writing out the numbers to represent the amounts when counting each set. The teacher reviews equal sets and counting items in each set to determine how many in all. The teacher shows students the hundreds chart or number line and invites students to suggest an amount to count by. To model counting, the teacher starts with zero and counts by ones to the amount students suggested she count by. The teacher then highlights or colors the number, then counts again and highlights the next number. An example is shown in Figure 3–15.

After counting the designated amount by ones and highlighting the numbers, the teacher models skip counting and touches the numbers she previously highlighted. She then writes an equation that represents what she did if students are ready to create equations. An example is shown in Figure 3–16.

Once the teacher models skip counting and writing an equation, the teacher invites students to count with her, providing them guidance to ensure success. After enough practice sessions in which the teacher provides guidance, she prompts students to count on their own. Based on student responses, the teacher provides feedback and questions that build understanding.

Number Line With Objects and Counting

Place an object next to each numeral on the number line.

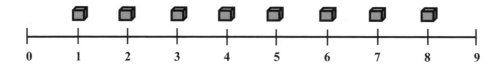

Think Aloud: I have objects next to the numbers. I am counting by two, so I hop up the number line two times. Here I go, *two.* Is that right? Are there two blocks? Yes. Next, *four, six.* Is that right? Are there six blocks? Yes. OK, next, *eight.* Is that right? Yes, there are eight blocks.

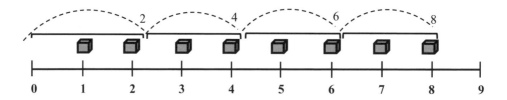

Ten Frame With Drawings

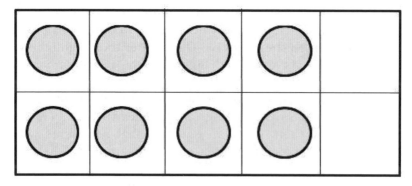

Figure 3–14. Representational Level Models for Skip Counting.

Number Line

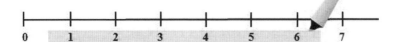

Hundreds Chart

1	2	3	4	5	6	7	8	9	10
11	12	13	14	15	16	17	18	19	20
21	22	23	24	25	26	27	28	29	30
31	32	33	34	35	36	37	38	39	40
41	42	43	44	45	46	47	48	49	50
51	52	53	54	55	56	57	58	59	60
61	62	63	64	65	66	67	68	69	70
71	72	73	74	75	76	77	78	79	80
81	82	83	84	85	86	87	88	89	90
91	92	93	94	95	96	97	98	99	100

Figure 3–15. Counting by Ones Highlighting Numbers.

1	2	3	4	5	6	7	8	9	10
11	12	13	14	15	16	17	18	19	20
21	22	23	24	25	26	27	28	29	30
31	32	33	34	35	36	37	38	39	40
41	42	43	44	45	46	47	48	49	50
51	52	53	54	55	56	57	58	59	60
61	62	63	64	65	66	67	68	69	70
71	72	73	74	75	76	77	78	79	80
81	82	83	84	85	86	87	88	89	90
91	92	93	94	95	96	97	98	99	100

$$2 + 2 + 2 + 2 + 2 = 10$$

Figure 3–16. Skip Counting Numbers Highlighted and Writing Equation.

CRA/CSA APPLICATION AND COUNTING QUANTITIES LARGER THAN 20

It is possible for students to learn mathematic procedures to solve problems, yet not truly understand the mathematical concepts behind the procedure (Fuson, 1998). It is also possible that students rely on procedures to solve problems and, at the same time, do not have a flexibility with numbers to solve problems in a fluid manner. Therefore, for students in primary grades and students who struggle with mathematics in upper elementary grades, instruction needs to revisit counting as a way of building flexibility with numbers.

Concrete Phase Counting Quantities Larger Than 20

In the beginning, the teacher facilitates a discussion about numbers being large enough that counting every item can become very long and tedious. To illustrate the point, the teacher lays out a quantity of 30 or more objects. She encourages the class to count each item with her to determine how many. She then explains that she will sort the objects into groups to make counting easier. She asks students to make suggestions for an amount to assign to each equal group. Usually, students provide small numbers such as five or three. With practice, students will start assigning larger numbers to create the equal sets. The teacher models sorting the amount into equal sets the students have suggested and skip counts the sets. Any remaining numbers that cannot be made into the defined sets, the teacher will use shortened counting to determine the total amount. The teacher also writes out an equation that represents adding each set together, which is repeated addition. An example is shown in Figure 3–17.

After modeling, the teacher lays out another quantity of objects of 20 or more and again holds a discussion of how many to assign to each set. She then invites students to count with her. Once the total is determined, she helps students write out an equation that represents counting each set to determine a total. After several practice sessions in which the teacher helps students count equal sets through skip counting, the teacher directs students to skip count quantities on their own. She distributes objects to each student and prompts students to create equal sets, skip count to determine the total, and write out equations that represent what they did. She monitors students and provides feedback and questions that scaffold understanding..

Representational/Semi-Concrete Phase Counting Quanties Larger Than 20

In students have ample practice skip counting sets of objects, instruction moves to the representational semi-concrete phase. In this phase of instruc-

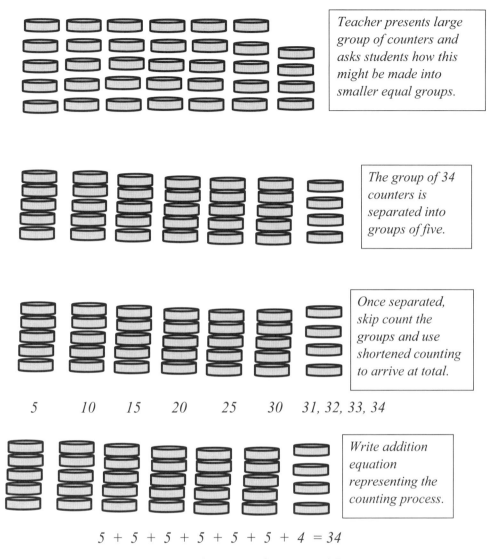

Figure 3–17. Sets of Objects With Shortened Counting and Equation.

tion, students can create drawings in journals or the teacher can use pictures that represent quantities larger than 20. An example is shown in Figure 3–18.

First the teacher reviews why people skip count and explains that skip counting is an easier way to count larger quantities. She shows students a journal or a picture with nonsymbolic amounts larger than 20 and models circling equal groups. She then skip counts the groups to determine the total amount and writes an equation that represents what she did. Once she models skip counting, she asks students to count quantities with her. She displays another amount larger than 20 using a journal or picture and

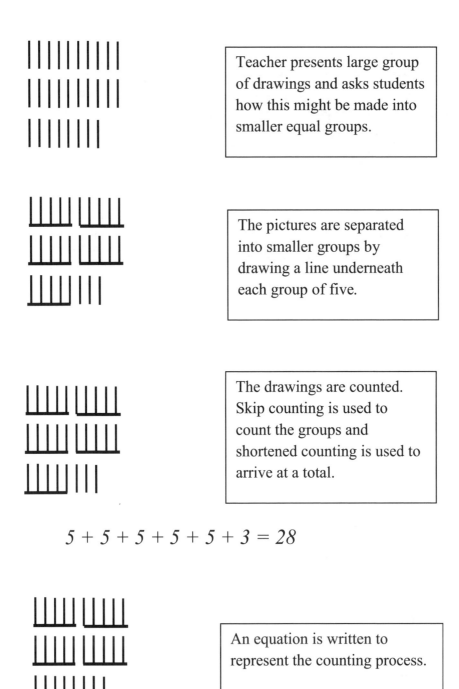

Figure 3–18. *Using Pictures to Count Quantities Larger Than 20.*

invites students to suggest an amount to designate to each set. Using the amount students suggest, she circles the amounts to create sets. The teacher and students skip count together to determine the total, and the teacher solicits suggestions for equations. After several practice sessions, the teacher directs students to skip count amounts independently. She provides students with journal questions that have quantities larger than 20 displayed or pictures that display amounts larger than 20. She prompts students to create sets, skip count, and write out equations that represent what they did. The teacher provides feedback based on students' response through suggestions and guided questions.

Abstract Phase Shortened Counting Quantities Larger Than 20

Once students have enough practice skip counting sets using pictures, instruction moves to the abstract phase. In this phase of instruction, students can use a number line or hundreds chart. An example is shown in Figure 3–19.

Once again, the teacher reviews why people skip count. She shows students a number line or hundreds chart. She asks students to suggest an amount to skip count. Using the amount students suggested, the teacher starts with zero and counts by ones using the number line or hundreds chart. The number she lands on after counting, she highlights or colors. She then counts the designated amount by ones again and highlights the next number. She continues at least three more times. An example is shown in Figure 3–20.

After counting and highlighting, the teacher models writing an equation that represents what she did. The teacher models skip counting another quantity, except this time counts by the designated amount instead of ones. She continues to highlight the number and writes an equation that shows what she did. An example is shown in Figure 3–21.

After the teacher models skip counting by the designated amount, the teacher asks students to join her. The teacher and students can choose another amount to skip count, and at first they count the amount together by ones, highlight the number, and then count again. The teacher prompts students to write an equation and provides assistance to those who need it. Once students have enough practice with scaffolding instruction, the teacher invites students to skip count using the number line and hundreds chart independently. The teacher provides feedback based on student responses through guided questioning. Once students master skip counting starting with zero, the teacher can create more of a challenge by having students skip count starting with a number instead of zero. For older students, the teacher can have students skip count with numbers greater than 100. This is very helpful for students who do not have a complete understanding of place value.

Counting by Five

1	2	3	4	5	6	7	8	9	10
11	12	13	14	15	16	17	18	19	20
21	22	23	24	25	26	27	28	29	30
31	32	33	34	35	36	37	38	39	40
41	42	43	44	45	46	47	48	49	50
51	52	53	54	55	56	57	58	59	60
61	62	63	64	65	66	67	68	69	70
71	72	73	74	75	76	77	78	79	80
81	82	83	84	85	86	87	88	89	90
91	92	93	94	95	96	97	98	99	100

Counting by Three

1	2	3	4	5	6	7	8	9	10
11	12	13	14	15	16	17	18	19	20
21	22	23	24	25	26	27	28	29	30
31	32	33	34	35	36	37	38	39	40
41	42	43	44	45	46	47	48	49	50
51	52	53	54	55	56	57	58	59	60
61	62	63	64	65	66	67	68	69	70
71	72	73	74	75	76	77	78	79	80
81	82	83	84	85	86	87	88	89	90
91	92	93	94	95	96	97	98	99	100

Figure 3–19. Hundreds Chart Used to Teach Skip and Shortened Counting.

CHAPTER SUMMARY

CRA/CSA is a way for students to build relevant and meaningful counting experiences because it allows students to literally manipulate and problem solve quantities found in their environment. Counting instruction for students at risk for mathematics failure involves the use of nonsymbolic representations of quantities with objects and then drawings, all while counting and enumerating amounts using symbolic representations of the number names or number symbols. The representational phase of CRA/CSA instruction

1	2	3	4	5	6	7	8	9	10
11	12	13	14	15	16	17	18	19	20
21	22	23	24	25	26	27	28	29	30
31	32	33	34	35	36	37	38	39	40
41	42	43	44	45	46	47	48	49	50
51	52	53	54	55	56	57	58	59	60
61	62	63	64	65	66	67	68	69	70
71	72	73	74	75	76	77	78	79	80
81	82	83	84	85	86	87	88	89	90
91	92	93	94	95	96	97	98	99	100

Figure 3–20. Counting by Ones.

1	2	3	4	5	6	7	8	9	10
11	12	13	14	15	16	17	18	19	20
21	22	23	24	25	26	27	28	29	30
31	32	33	34	35	36	37	38	39	40
41	42	43	44	45	46	47	48	49	50
51	52	53	54	55	56	57	58	59	60
61	62	63	64	65	66	67	68	69	70
71	72	73	74	75	76	77	78	79	80
81	82	83	84	85	86	87	88	89	90
91	92	93	94	95	96	97	98	99	100

$$10, 20, 30 \ldots 31, 32, 33, 34$$

$$10 + 10 + 10 + 1 + 1 + 1 + 1 = 34$$

Figure 3–21. Skip Counting and Shortened Counting With Equation.

allows for students to build understanding of symbolic numbers because it bridges understanding of nonsymbolic amounts with symbolic representations. It is important to have counting lessons build on each other and become more complex by including larger quantities as students develop conceptualizations of numbers.

Even though counting and cardinality are skill sets expected for younger students, it is possible for students to learn mathematic procedures to solve problems, yet not truly understand the mathematical concepts behind the procedure (Fuson, 1998). This is possible because students can rely on procedures to solve problems without having flexibility with numbers to solve problems in a fluid manner. Counting instruction will need to be revisited for students in primary grades and students who struggle with mathematics in upper elementary grades as a way of building flexibility with numbers.

APPLICATION QUESTIONS

1. Describe the stages of counting and how each stage relates to the principles of counting.

2. Provide reasons and describe how a teacher would implement ten frames or a number line during counting instruction.

3. Within the context of teaching students counting numbers higher than 20, describe each of the following: concrete instruction, representational instruction, and abstract instruction.

4. Within the context of teaching students skip counting, describe each of the following: concrete instruction, representational instruction, and abstract instruction.

5. Provide reasons and describe how a teacher would implement a hundreds chart during skip counting instruction.

REFERENCES

Clements, D. H. (1999). Subitizing: What is it? Why teach it? *Teaching Children Mathematics, 5,* 400–405.

Clements, D. H., & Sarama, J. (2010). Learning trajectories in early mathematics: Sequences of acquisition and teaching. In *Encyclopedia on early childhood development.* Center of Excellence for Early Childhood Development. Retrieved from http://www.child-encyclopedia.com/numeracy/according-experts/learning-trajectories-early-mathematics-sequences-acquisition-and

Common Core State Standards Initiative (CCSSI). (2010). *Common Core State Standards for Mathematics.* Washington, DC: National Governors Association Center for Best Practices and the Council of Chief State School Officers. Retrieved from http://www.corestandards.org/assets/CCSSI_Math%20Standards.pdf

Fuson, K. (1998). Pedagogical, mathematical, and real-world conceptual-support nets: A model for building children's multidigit domain knowledge. *Mathematical Cognition, 4,* 147–186.

Jordan, N. C., Kaplan, D., Locuniak, M. N., & Ramineni, C. (2007). Predicting first-grade math achievement from developmental number sense trajectories. *Learning Disabilities Research & Practice, 22,* 36–46.

Kolkman, M. E., Kroesbergen, E. H., & Leseman P. P. M. (2013). Early numerical development and the role of non-symbolic and symbolic skills. *Learning and Instruction*, *25*, 95–103.

Kroesbergen, E. H., Noordende, J. E. van't, & Kolkman, M. E. (2014). Training working memory in kindergarten children: Effects on working memory and early numeracy. *Child Neuropsychology*, *20*, 23–37.

Kroesbergen E. H., & Van Luit J. E. H. (2003). Mathematics interventions for children with special educational needs a meta-analysis. *Remedial and Special Education*, *24*, 97–114.

Mazzocco, M. M., & Thomspon, R. E. (2005). Kindergarten predictors of math learning disability. *Learning Disabilities Research & Practice*, *20*, 142–155.

Miller, S. P. (2009). *Validated practices for teaching students with diverse needs and abilities* (2nd ed.). Upper Saddle River, NJ: Pearson.

National Association for the Education of Young Children. (2009a). *Developmentally appropriate practice in early childhood programs serving children from birth through age 8*. A position statement of the National Association for the Education of Young Children. Retrieved from http://www.naeyc.org/files/naeyc/file/positions/PSDAP.pdf

National Association for the Education of Young Children. (2009b). *Ten effective developmentally appropriate practice teaching strategies*. Retrieved from http://www.naeyc.org/dap/10-effective-dap-teaching-strategies

Peterson, S. K., Mercer, C. D., & O'Shea, L. (1988). Teaching learning disabled students place value using the concrete to abstract sequence. *Learning Disabilities Research*, *4*, 52–56.

Powell, S. R., Fuchs, L. S., & Fuchs, D. (2013). Reaching the mountaintop: Addressing common core standards in mathematics for students with mathematics difficulties. *Learning Disabilities Research and Practice*, *28*, 38–48.

Van Luit, J. E. H., & Schopman, E. A. M. (2000). Improving early numeracy of young children with special educational needs. *Remedial and Special Education*, *21*, 27–40. http://dx.doi.org/10.1177/074193250002100105

CHAPTER 4

Teaching Addition Using the Concrete-Representational/ Semi-Concrete–Abstract Sequence

OVERVIEW

This chapter shows how the concrete-representational/semi-concrete–abstract sequence (CRA/CSA) is used to teach addition, from basic facts to addition of larger numbers that require regrouping. Instruction related to the development of understanding the addition operation will be shown through examples at the concrete, representational/semi-concrete, and abstract levels. The Standards for Mathematical Practice emphasize flexibility in thinking and problem solving (Common Core State Standards Initiative [CCSSI], 2010); therefore, the CRA/CSA application will be described and shown within different approaches to addition, which involve the decomposition of numbers, application of mathematical properties, and the traditional algorithm. This chapter will provide readers with a rationale for using methods and strategies to use the CRA/CSA sequence to teach students who struggle in mathematics as well as examples and guidelines for implementation.

SEQUENCE OF ADDITION INSTRUCTION WITHIN MATHEMATICS STANDARDS

Instruction related to addition begins in kindergarten and spans the elementary years. It begins without equations as students recognize the composition of numbers. Next, students learn the symbolic language of mathematics and use concrete models and drawings to solve to represent and solve problems. At each level of complexity in the addition operation, the standards begin at the concrete level and gradually require explanation, and fluent computation using numbers only. The standards (CCSSI, 2010) related to addition are located in Table 4–1.

Table 4–1. Mathematics Standards Related to Addition

Compose and decompose numbers from 11 to 19 into ten ones and some further ones (e.g., by using objects or drawings).

Record each composition or decomposition of numbers from 11 to 19 by a drawing or by using an equation (such as 18 = 10 + 8).

Demonstrate understanding that numbers 11 to 19 are composed of ten ones and one, two, three, four, five, six, seven, eight, or nine ones.

Use concrete models and drawings to add within 100 using strategies based on place value, properties of operations, and the relationship between addition and subtraction.

Given a two-digit number, mentally find 10 more than the given number without counting.

Using numbers only, fluently add and subtract within 100 using strategies based on place value, properties of operations, and/or the relationship between addition and subtraction.

Add up to four two-digit numbers using strategies based on place value and properties of operations.

Use concrete models and drawings to add within 1,000 using strategies based on place value, properties of operations, and the relationship between addition and subtraction.

Mentally add 10 or 100 to a given number 100 to 900.

Explain why addition strategies work, using place value and the properties of operations.

Using numbers only, fluently add within 1,000 using strategies and algorithms based on place value, properties of operations, and/or the relationship between addition and subtraction.

Fluently add multidigit whole numbers using the standard/traditional algorithm.

DESCRIPTION OF ADDITION AND PREREQUISITE SKILLS

Addition, at its most basic conceptual level, is joining. Conceptual understanding of addition begins before numbers are introduced within situations in which objects, toys, food items, people, and so on are combined to form another group. Young children's daily experiences involve the concept of addition when they combine blocks for building, bring dolls or stuffed animals together for a "party," or push their snack items together into one pile. The concept of joining is quantified and communicated using symbols by the addition operation within mathematics.

The prerequisite skills for execution of the addition operation begin with concepts related to numbers and counting (Witzel, Ferguson, & Mink, 2012). Prerequisite skills related to numbers are discussed in the first two chapters. Among these, rote counting involves knowing the order of numbers, saying number names in order. Students learn that each number symbol represents a particular amount; the numeral 5 represents a specific amount of objects that can be arranged in various ways. Another prerequisite skill is one-to-one correspondence in which a child can count a group of objects by assigning a number to each object, one count per object.

Development of conceptual understanding of addition begins with joining concrete objects, making meaning, and discussing the meaning of the new concrete representation of a number without mathematical symbols. When given a group of objects, a child can count how many are in the group and can explore different ways to combine the objects to form the whole group. For example, if given six objects, the students can "make six" with (a) two groups of three, (b) a group of four and a group of two, (c) a group of five and a group of one, or (d) a group of six and a group of zero. This concept of combining numbers to form another is made more formal through the use of numerals and symbols. Students learn the language of mathematics with the symbols that represent addition (+) and equal (=). The symbol, +, means that groups are joined and the symbol, =, means that the amount of objects on the right side of the symbol is the same as the amount of objects on the left side of the symbol. During this time, students also explore properties such as the commutative property and identity properties in addition. It is important that students recognize that when given addends, one will determine the same sum, regardless of which addend is counted first. They also explore how when zero is added to an addend, the sum is the same as the addend. Finally, as they work toward fluency in addition, students move from counting every object to utilizing patterns such as doubles, partners that make ten, and doubles plus one. As students explore patterns in addition, they also are encouraged to develop skills of counting on from the first addend used, rather than counting every single item represented in the addition sentence.

Advancement in understanding of addition requires understanding of numbers, place value, and the base 10 system (National Mathematics Advisory Panel, 2008; Witzel et al., 2012). Students must understand that the organization of our number system involves grouping numbers by tens. After counting to 10, the next nine numbers are 10 and some more (11 is 10 and one more); after counting more than 10, the next nine numbers are 20 and some more (22 is 20 and two more). Ten tens are one hundred and ten hundreds are one thousand and so forth. Students must understand this system in order to add numbers larger than nine, especially when regrouping is involved. Many students can identify ones, tens, hundreds, and so on by labeling columns; however, this skill does not adequately demonstrate an understanding of the base 10 system that is required to combine large numbers. For example, the addition of 25 and 27 involves combining five ones, seven ones, two tens, and two tens. A student must recognize that this combination will result in 12 ones and four tens, which, according to the base 10 system, is five tens and two ones. Most mathematics standards do not require that students reach the answer, 52 (five tens and two ones), in one particular way; however, any of the approaches require that the students understand the base 10 system and the relations between the numbers within it. The following sections will demonstrate how CRA/CSA provides the conceptual and procedural knowledge required to demonstrate understanding and fluency in the operation of addition.

CRA/CSA APPLICATION IN DEVELOPING UNDERSTANDING OF BASIC ADDITION

Making Numbers

In developing students' flexibility in understanding and communicating about numbers as a prerequisite to addition as a formal operation, a common task or game is making numbers. Students are given a numeral and asked how this number can be represented in different ways (e.g., given five, a student would reason that two and three make five; four and one make five; two, two, and one make five; or five and zero make five). This simple activity, when presented using numbers only, may be challenging for students who struggle in mathematics. When teachers determine, through formative assessment, that a student is struggling with the open-ended task of making numbers, the CRA/CSA sequence can provide the scaffolding needed that leads to abstract mathematical thinking about the composition of numbers. At the concrete level, the teacher would use objects and visual aids: present five objects and squares drawn on paper, large enough to group objects together. The teacher would show how two different numbers can be combined to make five by separating the five objects, placing objects in each of two squares. The teacher would count the objects, showing that there are five and then divide the five objects across the two squares, placing three in one square and two in another. Student engagement would be maintained by asking him/her to participate by counting the objects in each square. The teacher would explain that five is made with the combination of three and two. Through the manipulation of objects, the concept of number composition is made more explicit and the language involved in the mathematical process can be clarified. The teacher and student would describe how five can be made in multiple ways. The teacher would complete this process of demonstration at the concrete level for the other combinations of numbers that make five (e.g., four and one). More than two squares should be used in order to demonstrate that more than two numbers can be combined to make another number (e.g., two, two, and one make five or three, one, and one make five).

The next step is guided practice at the concrete level in which the teacher and student take turns in the process of making numbers; the student would count the objects, the teacher places one group in a square and the student places another group in a square. Finally, the student would complete the process without the teachers' assistance. During each phase of instruction, it is important to allow the student to share what he/she can do and to scaffold the instruction based on his/her current thinking. Questioning and building on how students respond is essential. Instruction in making numbers through concrete-level instruction is shown in Figure 4–1.

At the representational/semi-concrete level, the same procedures would be used, but the teacher represents numbers with semi-concrete drawings. If making the number 5, the teacher draws five tallies, drawing different combinations of across empty squares. If there were two squares, the teacher would draw three in the first square and two tallies in the other square. The same instructional procedures are used in which the teacher demonstrates, then guides through a back-and-forth process, and finally asks the student to complete the task without assistance. It is important that students see the connection between the concrete and representational/semi-concrete stage. Asking questions to ensure students realize these are different representations of the same concept (addition) is critical at each phase of the CRA/CSA instruction. Instruction in making numbers through representational instruction is shown in Figure 4–2.

The abstract level would involve the use of numbers only. The teacher would ask the student to make five, presenting the number without providing objects or the visual aids in the form of squares to prompt the student's actions in composing the number. Making numbers at the abstract level is a task that students will encounter within their general education classroom, and explicit instruction using the CRA/CSA sequence provides scaffolding in developing their understanding the concept of number composition.

Making numbers seems simple and basic; however, a mathematical thinker is one who can represent numbers and approach problems in multiple ways (Dacey & Drew, 2012). The development of reasoning within early numeracy skills lays the foundation for approaching the operation of addition in multiple ways. For example, when given the problem 14 + 19, this problem may be solved without paper by thinking about the numbers' relation to 15, 20, or both. Understanding different ways to make 15 and 20 allows for the reasoning involved in solving 14 + 19 by solving using one of the following: (a) 15 + 19 − 1, (b) 14 + 20 − 1, and (c) 15 + 20 − 1 − 1.

Basic Addition

The early CRA/CSA research focused on basic operations (Mercer & Miller, 1992; Miller & Mercer, 1993). Within this research, the CRA/CSA sequence was used to model operations using manipulative objects, drawings, and

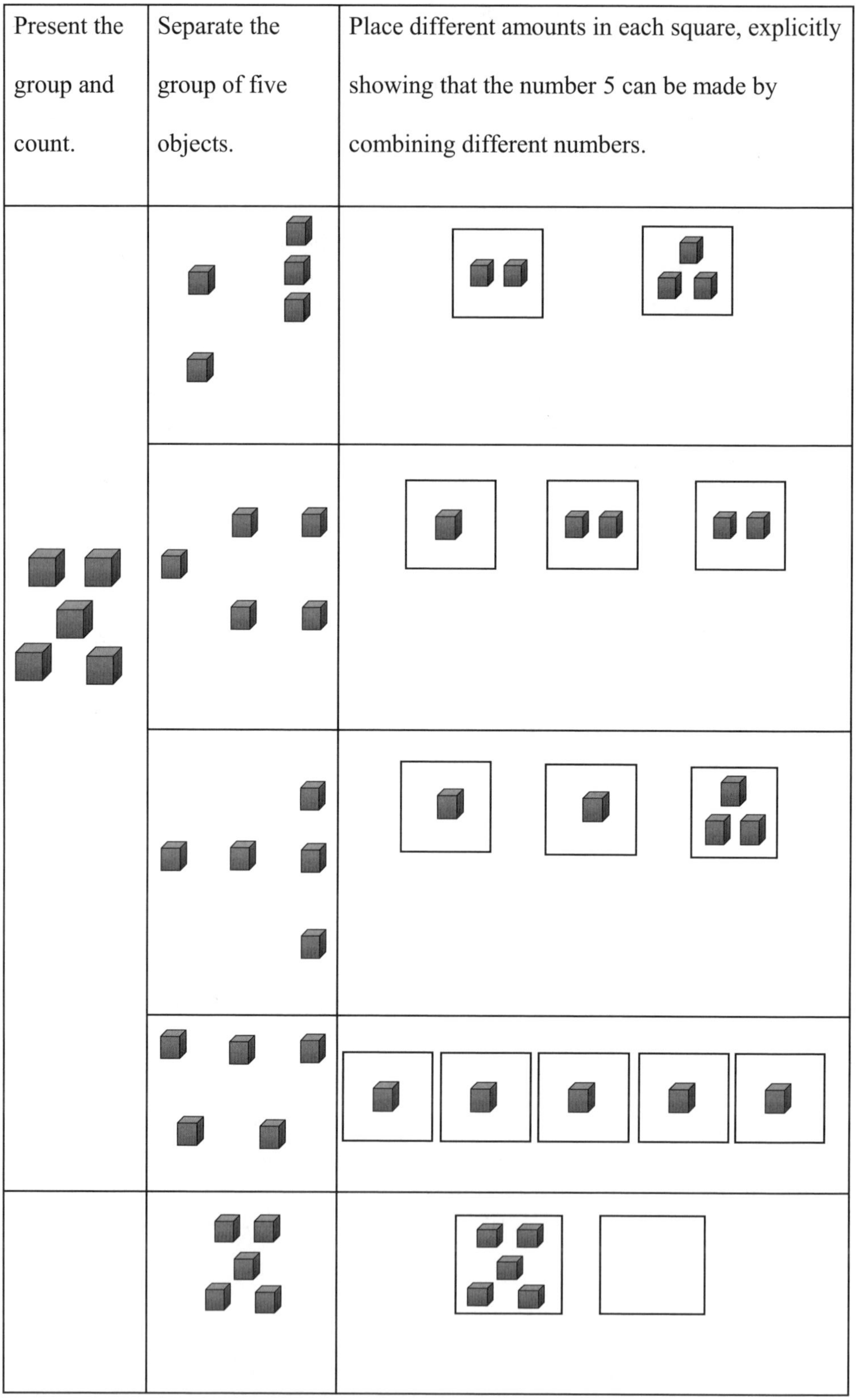

Present the group and count.	Separate the group of five objects.	Place different amounts in each square, explicitly showing that the number 5 can be made by combining different numbers.

Figure 4–1. *Making Numbers at the Concrete Level of Instruction.*

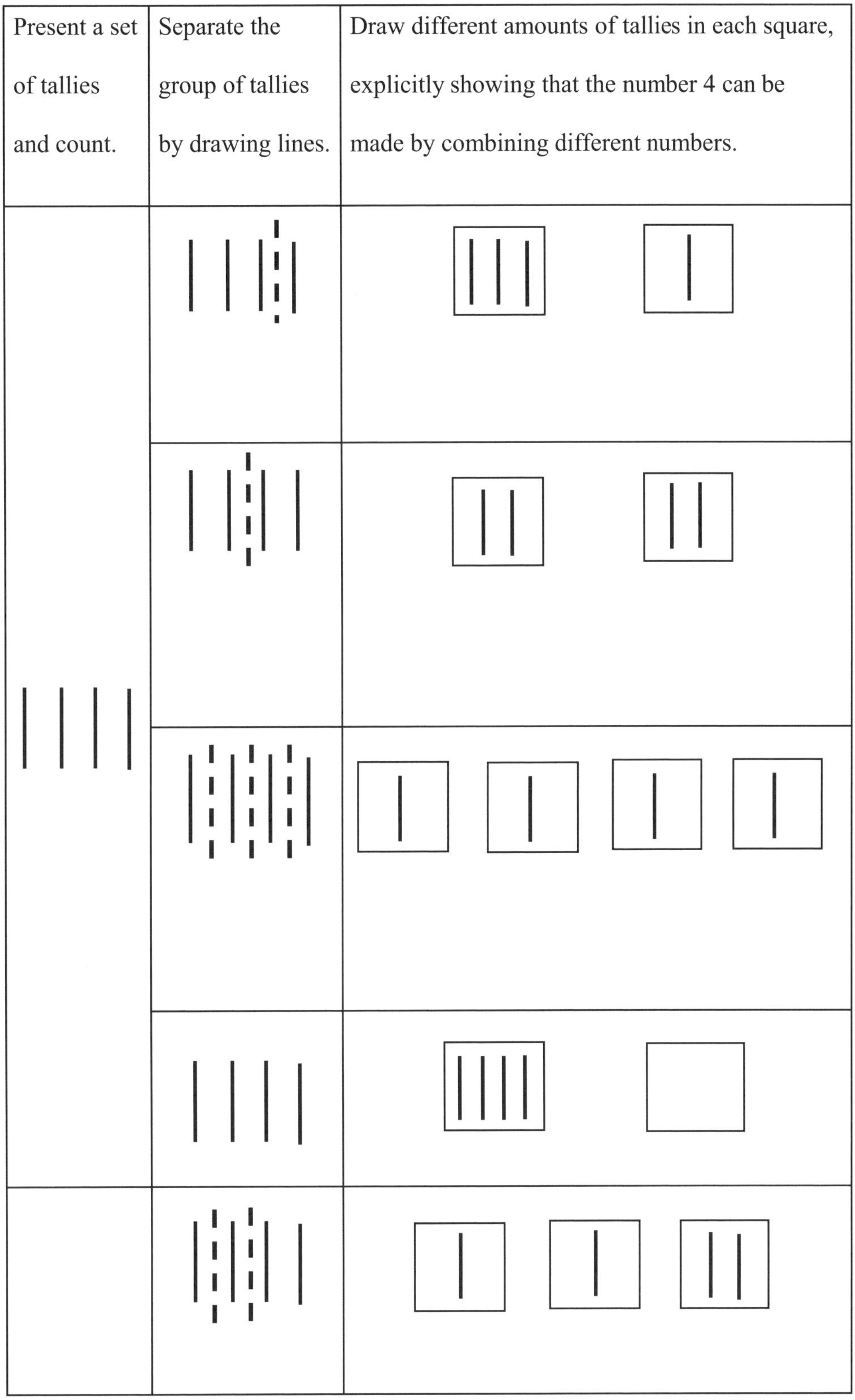

Present a set of tallies and count.	Separate the group of tallies by drawing lines.	Draw different amounts of tallies in each square, explicitly showing that the number 4 can be made by combining different numbers.
||||	|||¦|	||| |
	||¦||	|| ||
	¦|¦|¦|	| | | |
	||||	|||| [ ]
	¦|¦||	| | ||

Figure 4–2. *Making Numbers at the Representational Level of Instruction.*

numbers to develop conceptual understanding, procedural knowledge, and eventual fluency in addition facts. Although programs developed based on the research involve traditional approaches to operations, the CRA/CSA sequence applies to multiple approaches to solving problems. Mercer and Miller's research on CRA/CSA and basic addition involved combining groups of objects (concrete level) and drawings (representational, also known as semi-concrete) to form a sum. This approach allows for demonstration of equality, meaning that the amount of objects and drawings on one side of the equal sign is the same as the amount of objects and drawings on the other side of the equal sign. Understanding the meaning of the equal sign is essential, as many students have the misconception that the answer always comes after the equal sign (Mann, 2004). Students are able to explore and solve various addition problem types (such as join result unknown, join, change unknown, join start unknown, part-part-whole, whole unknown) contextually using objects (concrete level), drawings (representational or semi-concrete level), and eventually only abstract symbols.

As objects (concrete level) and drawings (representational or semi-concrete level) are manipulated, this physical process aids language development since the student can describe what he or she sees and physically manipulates. This develops a mental framework for communication of abstract processes in which addition problems are solved and discussed using just numbers and symbols. Basic addition should be taught within the context of real-life situations that students might experience on a regular basis. Simple word problems should be translated into equations, rather than only providing printed equations for students to solve. This ability to contextualize and decontextualize problems is a Standard of Mathematical Practice within the Standards of Mathematics (CCSSI, 2010). As students gain confidence and conceptual understanding, they can create problems to solve using equations which allows teachers to assess understanding as well. Beginning with initial instruction, the concept of joining or combining things together should be presented in multiple scenarios. Students who struggle with mathematics may not easily connect isolated computation to real situations, making the completion of word problems difficult. Therefore, the language associated with joining to form other quantities should be part of beginning and successive instruction. Initially joining and part-part-whole problems with the whole unknown are used, but early into instruction, other problem types are modeled to support the students' understanding of the equal sign and importance of context in problems.

Traditional Instruction

Teaching addition in a traditional approach using CRA/CSA involves translating the equation into words; the expression 3 + 2 is "three added to two." Concrete instruction shows the problem solving process in which three objects are joined with two objects to form another amount. Within this concrete process, the mathematical symbols for equal are explained as well. The symbol, =, as well as the horizontal line drawn underneath an equation that is written vertically both mean that the amounts on either side of these symbols are the same. We propose that once this is established, the other problem types, such as part-part-whole with the part unknown or join problems with the joined element being unknown, are introduced. The process for concrete basic addition instruction is shown in Figure 4–3. The first problem example is a join problem and the second problem is a part-part-whole example; within each problem, the result or whole is unknown.

The steps shown above are presented using explicit instruction. This means that a teacher would model and demonstrate the process by physically manipulating objects as well as thinking aloud. While demonstrating, the teacher ensures that the student remains engaged by asking the student to participate in steps that include previously mastered skills such as counting objects with the teacher or verbally repeating what the teacher says. After demonstration, the teacher guides the student by solving problems together. The student and the teacher take turns solving a problem, sharing thoughts

Present and translate the problem.	Representation With Manipulatives	Join and count the sum to arrive at an answer. Ensure that the amounts on either side of the equal symbol are the same.
Kiara had three pencils in her desk. She found two pencils in her backpack. How many pencils does Kiara have now? 3 + 2 = Kiara has three and two are added. Three plus two.	3 + 2 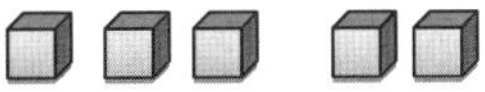 =	3 + 2 = 5 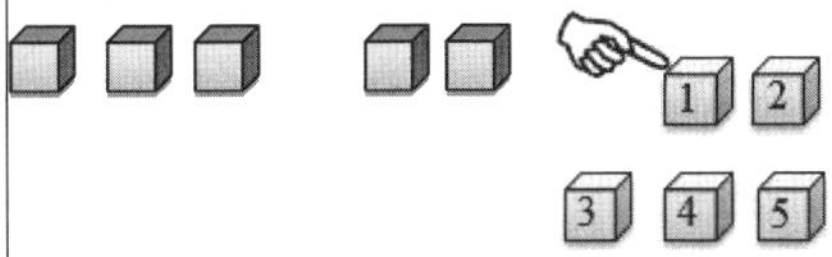
Ali ate four Oreos and three Vanilla Wafers for a snack. How many cookies did she eat in all? 4 + 3 Ali ate four cookies and then ate three more. Four plus three.	4 + 3 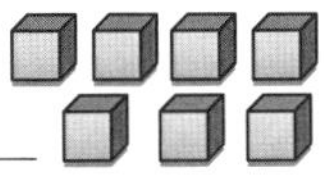	4 + 3 7 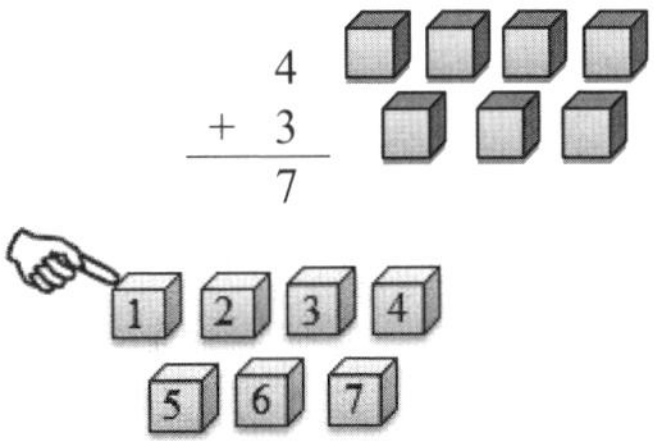

Figure 4–3. Addition Instruction at the Concrete Level.

about why they are doing each step, and the teacher provides prompts for the student verbally or visually by pointing. Finally, the teacher presents problems to the student for independent practice in which the student uses objects to solve addition problems without assistance and shares his or her own thinking aloud for the teacher to formatively assess.

After mastery of addition at the concrete level, the teacher will work to fade student dependence on objects by introducing the use of drawings. This is representational or semi-concrete level instruction in which the numbers within addition problems are represented using simple drawings. With a written equation (either created based on a simple word problem or preprinted), the amount represented by each number is written underneath or next to each. The total amount is then drawn after the equal symbol (either the symbol, =, written after a horizontally written equation or a line written underneath the vertical equation). The process for representational/semi-concrete level addition instruction is shown in Figure 4–4. The problems shown are part-part-whole with the whole unknown.

<table>
<tr><td>Present and translate the problem.</td><td>Representation With Manipulatives</td><td>Join and count the sum to arrive at an answer. Ensure that the amounts on either side of the equal symbol are the same.</td></tr>
<tr><td>Darian has two red markers and four blue markers. How many markers does he have in all?

4 + 2 =
Darian has red and blue markers. The different kinds of markers are combined.
Four plus two.</td><td>4 + 2 =
IIII II</td><td>4 + 2 = 6
IIII II
IIIIII
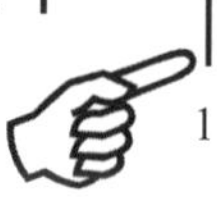
1 2 3 4 5 6</td></tr>
<tr><td>There are four boys and three girls on the playground. How many students are on the playground?

4
+ 3

The number of boys and girls is combined to find the total number of students.
Four plus three.</td><td>4 IIII
+ 3 III</td><td>4 IIII
+ 3 III
7
IIIIIII
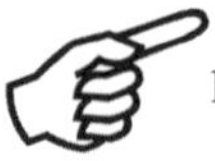
1 2 3 4 5 6 7</td></tr>
</table>

Figure 4–4. *Addition Instruction at the Representational Level.*

After mastery of addition using drawings, abstract instruction begins. Curriculum materials designed by Mercer and Miller (1992) include a mnemonic strategy for solving problems using numbers only: (a) discover the sign, (b) read the problem, (c) answer or draw and check, and (d) write the answer (DRAW). The DRAW strategy is included in all of Mercer and Miller's curriculum materials associated with basic operations and will be revisited across multiple chapters of this text (Mercer & Miller, 1992, 1994; Miller & Mercer, 1994). This mnemonic strategy serves as a bridge between representational and abstract computation. The DRAW strategy also provides students with procedures that will lead to accuracy in addition by (a) asking the student to attend to the operational symbol and the numbers within the equation, common errors made by students who struggle, and (b) prompting the student to draw the problem in situations in which the answer cannot be recalled from memory. During the abstract phase, students solve problems using the DRAW strategy, and the focus of instruction is fluency in addition facts. Since the concrete and representational/semi-concrete phases emphasize conceptual understanding of the operation as joining numbers, abstract-level instruction involves teaching students about rules and properties that will be used as students' memory or addition facts become automatic and students fluently compute problems. These rules emerge as students solve problems at the abstract level and the teacher scaffolds conversation such that the patterns emerge. These include (a) the zero rule (when zero is added to any number, its value is unchanged), (b) the one rule (one added to any number is one more; count up), and (c) the order rule or commutative property, which states that changing the order of numbers within the problem does not change the answer. The Standards for Mathematical Practice call for looking for and utilizing repeated reasoning. For example, asking the student to solve three to four problems where zero is an addend allows the teacher to then ask the students if they notice a pattern. Once the prediction is made that the value is unchanged, the students can test this pattern with a few more problems. Abstract-level instruction includes computation activities and games that will encourage quick and accurate addition. In order for students to advance to more complex levels of mathematics, basic facts must be automatic; this is defined as recall of the correct answer within 2 seconds of problem presentation (J. L. Hosp, M. K. Hosp, Howell, & Allison, 2014). This prevents excessive mental processing that detracts from a student's focus on more complex operations and mathematical tasks.

Counting On

Another approach to addition is counting on, and this can be taught using the CRA/CSA sequence as well. In contrast to the traditional approach to joining numbers in which two amounts are physically put together, counting on involves counting forward from one of the given numbers. When given the problem, 4 + 2, one counts forward two times from four: five . . . six. The difference between the traditional approach and counting on appears subtle, but it lays the foundation for mental computation used at the abstract level.

It is important that teachers are careful to ensure that students begin counting on with the number that follows the first number given in the problem. A frequent error of students who struggle with mathematics is to count beginning with the given number and arrive at an answer that is one less than it should be. Using the same example with 4 + 2, this error would be counting forward with four: four . . . five. Instruction at the concrete level will provide a foundation serving to prevent this type of error. The student can see the physical representation of the first number onto which the counting will begin and count forward to the next number with the objects representing the second number. The process for concrete counting-on addition instruction is shown in Figure 4–5.

Present and translate problem	Representation with Manipulatives	Join and count the sum to arrive at answer. Ensure that the amounts on either side of the equal symbol are the same
Kate has three cats and two dogs. How many pets does Katie have? 3 + 2 = Cats and dogs are pets Combine the number of cats and dogs to find the total number of pets Three plus two.	3 + 2  = ▢▢▢ ▢▢	3 + 2 = 5  ▢▢▢ ▢▢ ▢▢▢ [4][5][6]
Sam read three books last week and read three books this week. How many books did Sam read in all? 3 + 3 Sam read three books and then three more. The number of books are combined to find the total Three plus three.	3 + 3 ▢▢▢ ▢▢▢	3 + 3 ▔▔ 6 ▢▢▢ ▢▢▢ ▢▢▢ [4][5][6]

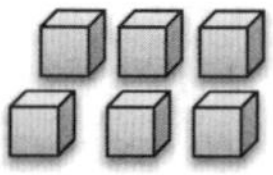

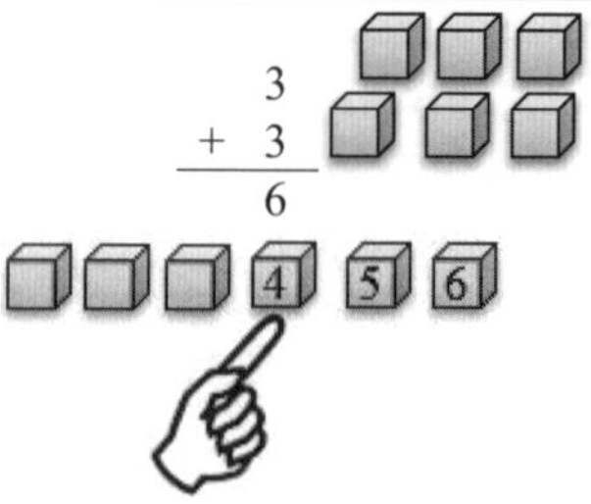

Figure 4–5. Concrete Level Addition Instruction Using Counting On.

4. Teaching Addition Using the Concrete-Representational/Semi-Concrete–Abstract Sequence **71**

Within representational/semi-concrete instruction, objects are replaced with simple drawings. The first number is represented using tallies, and the second number is added by drawing additional tallies, counting forward as they are drawn. The process for representational counting-on addition instruction is shown in Figure 4–6.

Another representational/semi-concrete method to assist students with counting on is the use of the number line. This provides another conceptualization of numbers using distance; however, a common error associated with number lines is to include the first addend in the counting-on process rather than the jumps from one number to another. If using this type of representation, it is critical that this distinction is clear. An example of the number line used for representational instruction is shown in Figure 4–7.

The focus of abstract instruction is automaticity and fluency. Although counting on provides a strategy on which a student can rely when memory

Present and translate the problem.	Representation With Manipulatives	Join and count the sum to arrive at an answer. Ensure that the amounts on either side of the equal symbol are the same.
Tim has three pieces of candy. Tim's sister gave him two more pieces of candy. How many candies does Tim have now? 3 + 2 = Tim had candy and two pieces were added to the amount he had. Three plus two.	3 + 2 = ||| ||	3 + 2 = 5 ||| || ||||| 4 5

There are four blue chairs and three orange chairs at the reading table. How many chairs are at the reading table?
 	
  4
+ 3
 	
The number of blue and orange chairs is combined to find the total number of chairs. Four plus three.
	4
	+ 3 ||||
	           |||
		4
		+ 3
		  7 ||||
		         |||
		
		|||||||
		     5 6 7

Figure 4–6. Representational Addition Instruction Using Counting On.

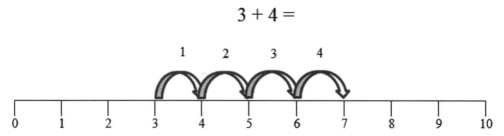

Figure 4–7. Number Line Used for Counting On Instruction at Representational Level.

fails, it is not preferable that a student mentally count on for each fact that is presented. Reliance on a counting-on strategy for recalling addition facts is more efficient than drawing or mentally counting every object. However, it is not as efficient as automatic recall of addition facts.

Missing Addend

When teaching basic addition, problem presentation should include varied presentation of problem types. To this point, the unknown number has been the sum. Students' understanding of the equal symbol is further conceptualized when presented with problems such as 2 + __ = 5. Since the amount on both sides of the equal symbol is the same, the missing addend can be found. It is critical that initial instruction in addition include explicit instruction in the concept of equal and the role of its symbol in mathematics. Missing addend problems may be presented as those in which a known amount and unknown amount are joined and a sum is known or a known part and an unknown part make up a known whole. Concrete-level instruction will provide the student with a physical model in which the student can observe both the given addend and the missing addend as composing the sum. Examples of concrete and representational instruction are shown in Figure 4–8.

Missing addend instruction at the abstract level, as discussed with previous approaches, involves numbers only. At this level, students should be moving toward automaticity. Activities and games associated with increasing recall should involve missing addends as well as missing sums. As students approach automaticity, a counting up might be helpful as long as the student uses this approach accurately by keeping track of the "counts" between the given addend and the sum.

Alternative Facts

The mathematical practices call for students to approach addition in multiple ways, and the CRA/CSA sequence can assist in this process. When given the problem nine plus four, nine objects would be combined with four objects and

Present and translate the problem.	Representation With Manipulatives	Separate the amount of the known addend from the whole. Notice the two numbers that make the whole. Place the second amount under the missing number. Ensure that the amounts on either side of the equal symbol are the same.
Cal and his brothers ate four cookies and some cupcakes for dessert. Altogether, they ate six desserts. How many cupcakes did they eat? 4 + __ = 6 Four cookies and some cupcakes were added together to make six desserts.	4 + __ = 6 	4 + 2 = 6 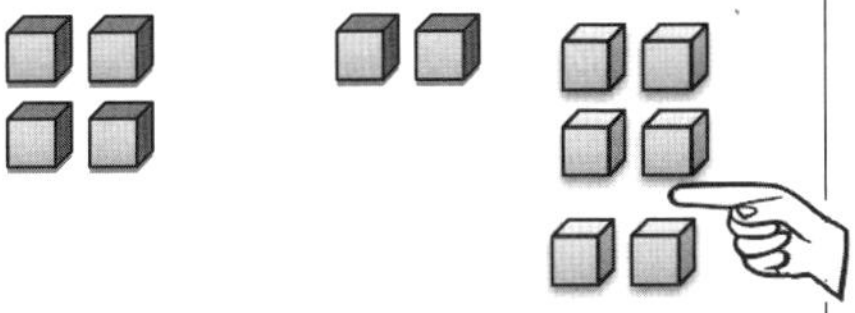

Raul had 4 dollars in his bank. Then he earned more money doing chores. Raul had 7 dollars total. How many dollars did he earn doing chores?
 	
   4
+ ?
―――
   7
 	
Raul had 4 dollars and earned more for a total of 7 dollars. Four plus an unknown number is 7.
	4
	+ ?
	―――
	   7 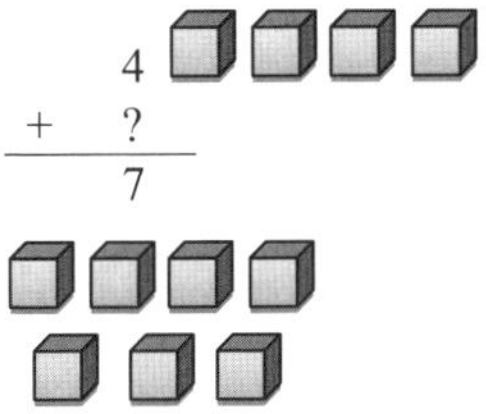
		4
		+ 3
		―――
		   7 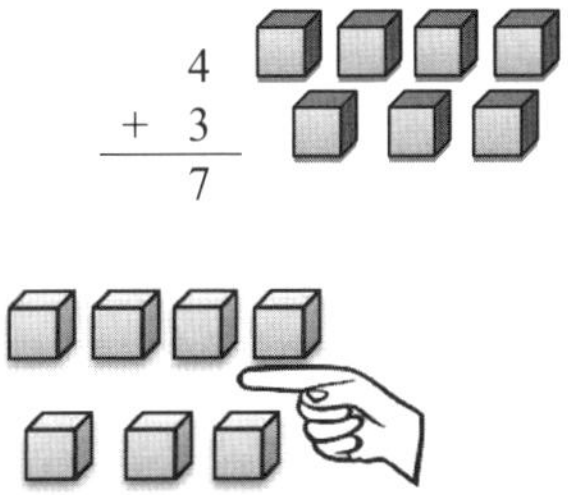

Figure 4–8. Missing Addend Instruction at Concrete and Representational Levels. continues

the total counted to arrive at 13. In further developing mathematical thinking and flexibility with the same problem, an alternative and perhaps easier problem can be generated when nine objects and four objects are shown. Using the two groups, move the objects so that the group of nine is now a group of 10; the combination of 10 and 3 is an easier problem. The process of changing 9 + 4 into 10 + 3 through the manipulation of each number (9 + 1 and 4 – 1) is a challenging mental process. Therefore, provide instruction at the concrete and representational/semi-concrete levels prior to asking students to mentally solve problems at the abstract level. Use of alternative facts can assist students as they attempt to become fluent in basic addition. This strategy can

Present and the translate problem.	Representation With Manipulatives	Circle the amount of the known addend within the total. Notice the two numbers that make the total. Count the amount that is not circled and write that under the missing number. Ensure that the amounts on either side of the equal symbol are the same.
Tyler had three pencils and his friend gave him some more. Now Tyler has five pencils. 3 + __ = 5 Tyler had three pencils and some were added to make 5. Three plus an unknown number is 5.	3 + __ = 5 ||| |||||	3 + __ = 5 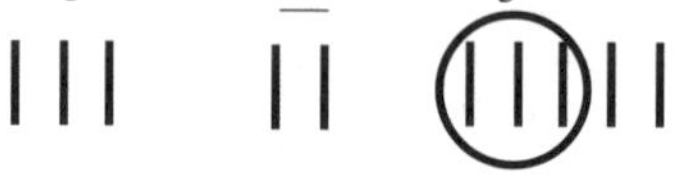 ||| || ⊙||||

Malik has four cousins who are girls. Malik has six cousins in all. How many of Malik's cousins are boys?
 	
   4
+ ?
―――
   6
 	
The four girls and the number of boys are added together for a total of six cousins. Four plus an unknown number is 6.
	4 ||||
	+ ?
	―――
	6
	
	||||||
		4 ||||
		+ 2 ||
		―――
		6
		
		⊙||||⊙||

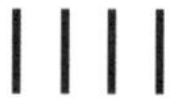

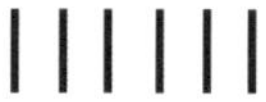

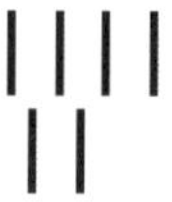

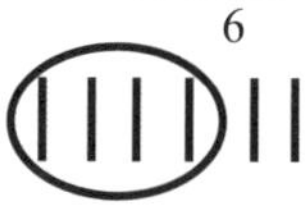

Figure 4–8. continued

also be used as students face more complex addition problems, allowing them to mentally solve problems involving larger numbers. For example, 29 + 41 becomes 30 + 40, a problem that does not require written computation. The processes for concrete and representational/semi-concrete addition instruction using alternative facts are shown in Figure 4–9.

CRA/CSA APPLICATION FOR ADDITION WITH REGROUPING

The use of the traditional algorithm is one way to approach addition with regrouping; however, it is not presented to students first since other approaches may assist them in understanding how the traditional algorithm

Problem: Sara had four pieces of candy. Tina had nine more pieces of candy than Sara. How many pieces of candy did Tina have?

Think Aloud: *The problem asks about the pieces of candy that Tina has. We do not know the amount, but we know that she has nine more than Sara. In the problem, the amounts are compared and we know that Sara has four and Tina has nine more than Sara's amount. Tina has the most amount of candy. To find the answer, we combine the amount Sara has with the comparison number, 9. We add four and nine.*

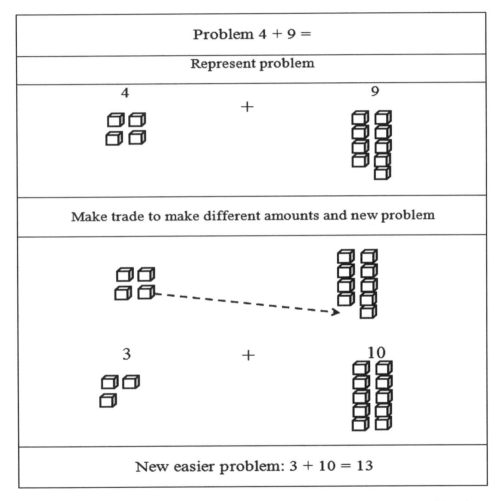

Figure 4–9. Addition Using Alternative Facts at Concrete and Representational Levels. continues

works. In supporting the development of mathematical thinking, students are taught to use their conceptual knowledge of numbers and operations to solve problems in a variety of ways (Barlow & Harmon, 2012). The CRA/CSA sequence can support multiple approaches beyond the traditional algorithm.

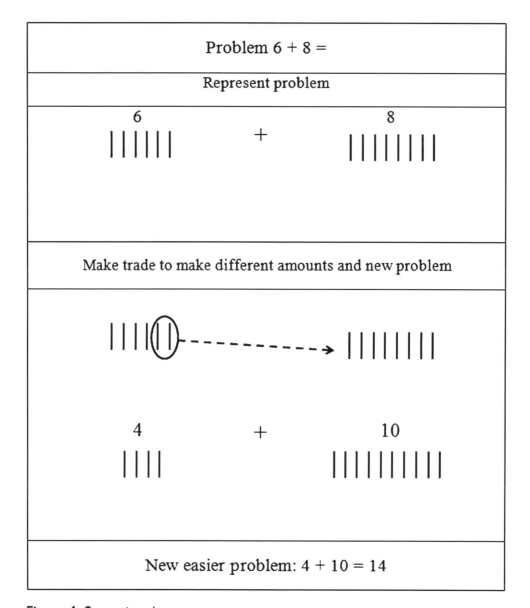

Figure 4–9. continued

When presented with addition with regrouping problems, these problems can be solved by decomposing each number, separating the number according to place value in columns and adding partial sums. A place value mat is recommended when working at the concrete level so that students can organize their base 10 blocks. Once numbers are represented using base 10 blocks, the numbers are added in each column. After adding the columns, those partial sums are combined to arrive at the result (e.g., 135 + 146 = 200 + 70 + 11 = 281). At the representational/semi-concrete level, problems are solved using drawings within a place value chart. Simple drawings represent ones (short tallies), tens (long vertical lines), and hundreds (squares). An example of using partial sums to solve an addition problem at the concrete and representational/semi-concrete level is shown in Figure 4–10.

Research has shown that CRA/CSA is effective in teaching students who struggle with mathematics addition with the traditional algorithm associated with regrouping (Miller & Kaffar, 2011). The CRA/CSA sequence supports the development of conceptual knowledge of the operation. Within the curriculum materials developed as a result of this research, problems are solved at the concrete level using base 10 blocks, which are organized using a place

Problem: There are two different choices on the school lunch menu, hamburgers and tacos. Three hundred students brought their lunch from home and the others chose to eat a school lunch. Of the students who ate at school lunch, 164 students chose to eat a hamburger and 147 students chose to eat tacos. How many students chose to eat a school lunch?

Think Aloud: *The problem asks about how many students ate a school lunch. There is information about students who brought their lunch and students who ate school lunches. We do not need to know about students with lunches from home, only those who ate a school lunch. There are two kinds of students who ate school lunch. Some ate hamburgers and others ate tacos. Altogether, they ate school lunch, but this number is unknown. This is a part-part-whole problem. We know the parts, hamburger and taco eaters, but not the whole. The parts are joined together to make the whole. When groups are joined, the operation is addition.*

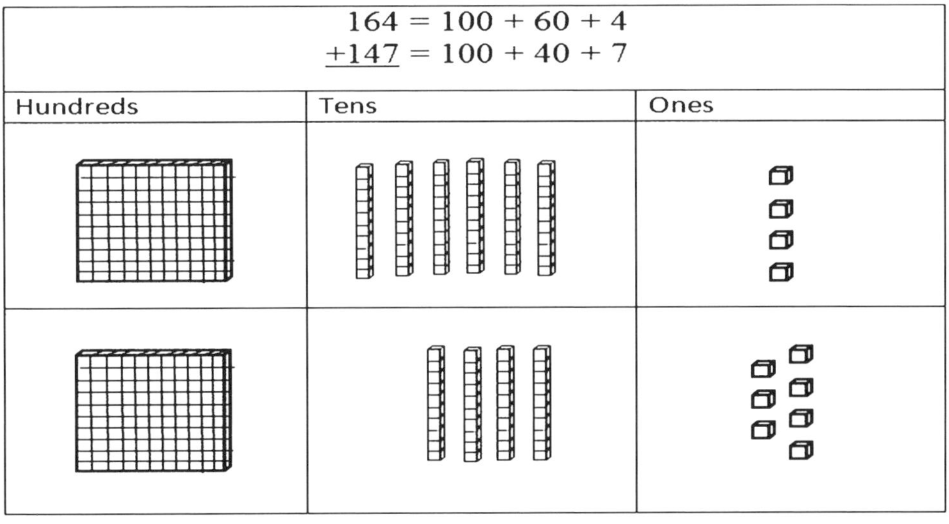

$$164 = 100 + 60 + 4$$
$$+147 = 100 + 40 + 7$$

Hundreds	Tens	Ones
[1 hundreds block]	[6 tens rods]	[4 ones cubes]
[1 hundreds block]	[4 tens rods]	[7 ones cubes]

Figure 4–10. Solving Addition With Regrouping Using Partial Sums at the Concrete and Representational Levels. continues

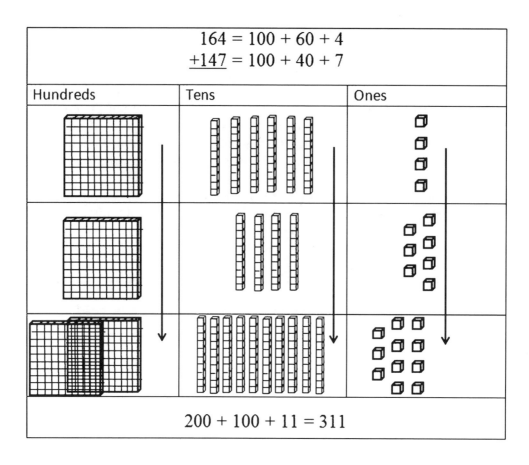

Figure 4–10. continues

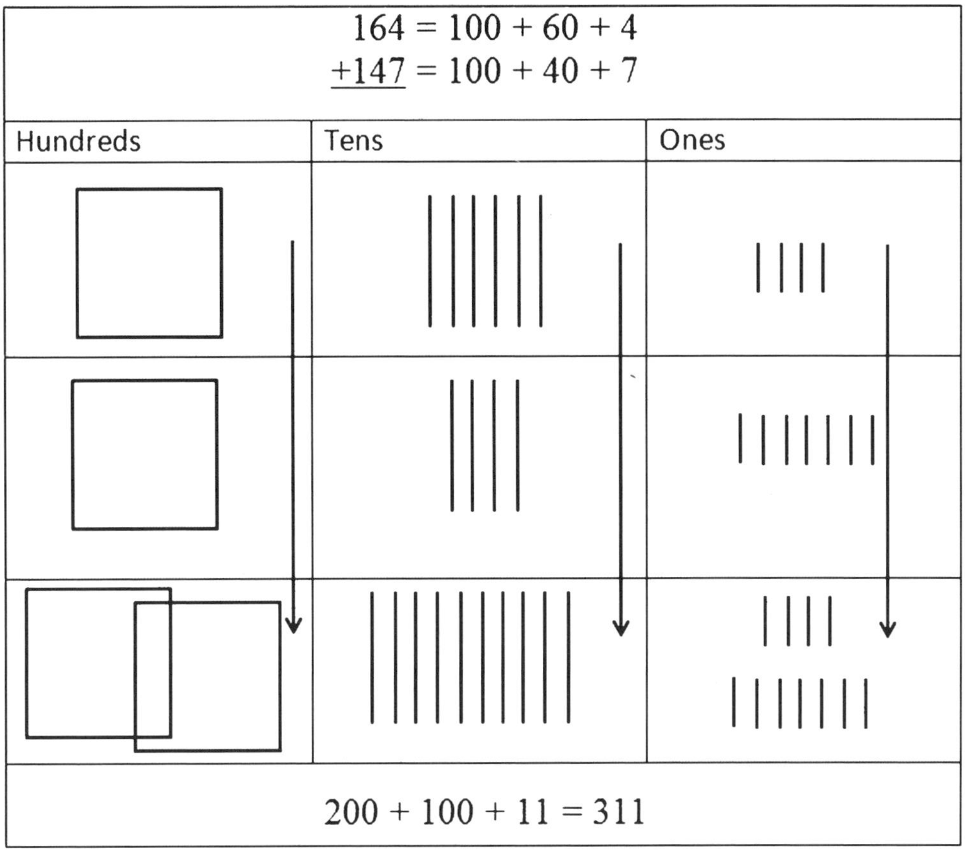

\begin{align} 164 &= 100 + 60 + 4 \\ +147 &= 100 + 40 + 7 \end{align}		
Hundreds	Tens	Ones
☐	||||||	||||
☐	||||	|||||||
☐☐	|||||||||||	|||| |||||||
200 + 100 + 11 = 311		

Figure 4–10. continued

value mat (Miller & Kaffar, & Mercer, 2011). The use of a place value mat supports students in organizing materials as well as the procedural components of traditional addition algorithm. Solving problems using manipulatives and a place value mat provides students with visual cues for the mathematical language used when explaining the computational process. When adding large numbers using the traditional algorithm, the regrouping processes are represented in a shortened form; students who struggle in mathematics may have weak number sense and lose sight of the meaning of each number, focusing on each individual numeral and on procedures. Instruction at the concrete level shows each number in the form of base 10 blocks and the process by which 10 ones are traded for 10 and why the resulting 10 is added to the tens column. Concrete-level instruction also provides students with the opportunity to observe the physical changes in numbers and participate in the process so that the language associated with the algorithm and mathematical procedures is apparent and tangible. It is important that both teachers and students communicate the actual number values throughout

80 *Making Mathematics Accessible for Elementary Students Who Struggle*

each step. An example of concrete instruction in addition with regrouping is shown in Figure 4–11.

Concrete-level instruction focuses on conceptual understanding of the addition operation as large numbers are combined and regrouping is required. In order to continue internalizing conceptual understanding but also fade dependence on objects, representational/semi-concrete instruction involves drawing problems using a place value chart. This is a smaller form of the place value mat used to organize base 10 blocks. An example of representational instruction in addition with regrouping is shown in Figure 4–12.

Presentation of Problem and Teacher's *Think Aloud*

At the football game, the fans could buy drinks and popcorn at the snack stand. In total, 234 cups of soda and 187 boxes of popcorn were sold. How many cups and boxes were used at the snack stand?

The problem asks about the number of cups and boxes used. The drinks sold were in cups and the popcorn was sold in boxes. To find the answer, we will combine cups and boxes. When we combine or join amounts, the operation is addition. So, we will add 234 to 187.

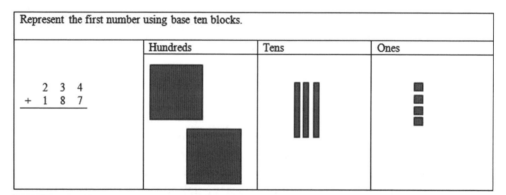

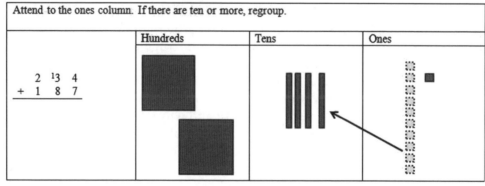

Figure 4–11. *Addition With Regrouping Instruction at Concrete Level.* continues

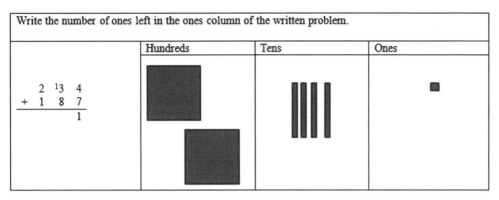

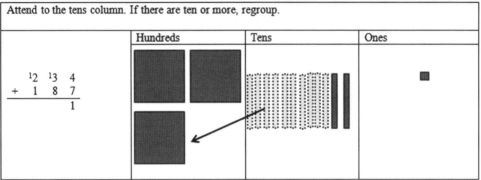

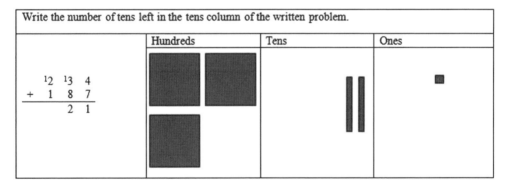

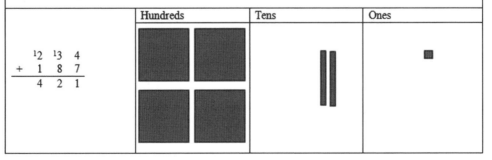

Figure 4–11. continued

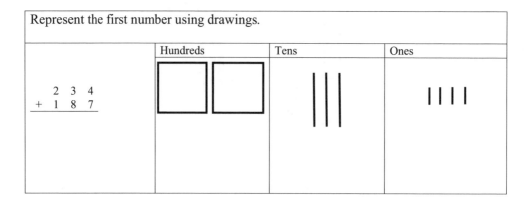

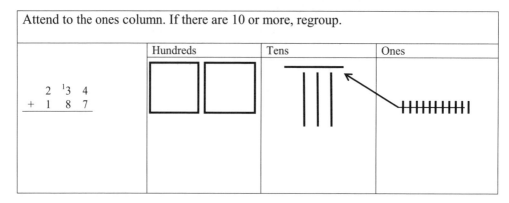

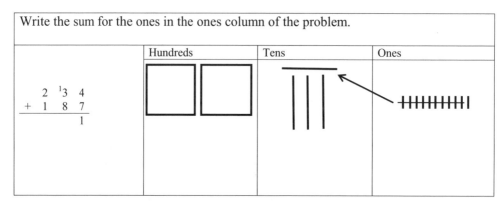

Figure 4–12. *Addition With Regrouping Instruction at Representational Level.* continues

The traditional addition algorithm involves procedural knowledge, which students use when solving problems using just numbers at the abstract level. It is critical that students have conceptual knowledge of this complex operation before there is a focus on procedural knowledge. This is a common problem for students who struggle in mathematics; they have poor conceptual knowl-

4. Teaching Addition Using the Concrete-Representational/Semi-Concrete–Abstract Sequence 83

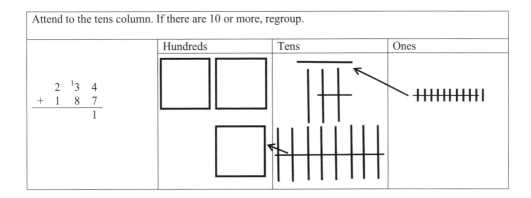

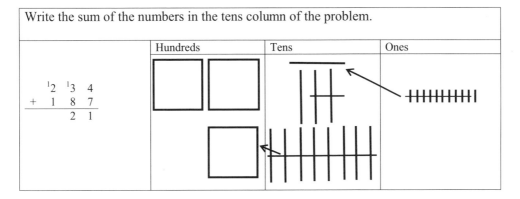

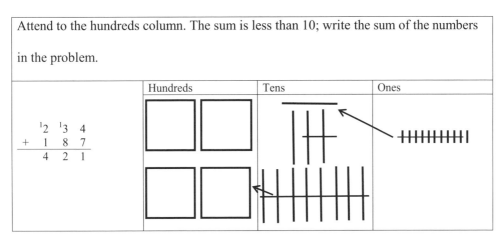

Figure 4–12. continued

edge, so they are left to follow a set of meaningless procedures, which then lead them to error patterns and confusion. The concrete manipulatives and discussion with the teacher can help students understand what is occurring when the algorithmic procedures are used at the abstract level.

Miller and Kaffar (2011) developed a mnemonic strategy to be used during the abstract level of instruction. The strategy provides an efficient set of steps, applicable to any addition with regrouping problem. The strategy is taught and used with numbers only, after conceptual understanding has been mastered using base 10 blocks (concrete level) and drawings (representational/semi-concrete). Miller and Kaffar's procedural strategy is a follows: (a) read the problem, (b) examine the ones column (use the *Ten or More* sentence; if there are 10 or more, go next door), (c) note ones in the ones column, (d) address the tens column (use the *Ten or More* sentence), (e) mark tens in the tens column, and (f) examine and note hundreds, and exit with a quick check. Figure 4–10 shows application of the RENAME strategy in solving an addition with regrouping problem at the abstract level. While this device can be useful in remembering the steps, it should only be used once students understand the reasons and meaning behind each step. Memorizing these rules without understanding can lead to future problems. The RENAME strategy is presented in Figure 4–13.

Another way to solve an addition problem that requires regrouping is by using alternative numbers. This approach involves changing addends by trading numbers, transforming a problem into one that is easier to solve. It is likely that this easier problem can be solved mentally rather than on paper if the student has strong flexibility with numbers. This is an example of solving an addition with regrouping problem by moving numbers: 156 + 118 = 160 + 114 = 274. However, for students who do not have flexible thinking at this level, modeling what is occurring at the concrete and representational levels can be useful. It also might be a strategy they prefer to use once they understand it. For instance, this type of thinking can help a student solve 199 + 199 in a way that is much easier and less mentally and physically taxing. Instead, the student considers an alternative problem, 200 + 198, rather than having to add and regroup as much as would be required with the traditional algorithm. An example of using alternative numbers is shown in Figure 4–14.

Problem	Strategy	Think aloud
$\begin{array}{r} 11 \\ 146 \\ +155 \\ \hline 301 \end{array}$	**R**ead the problem	The problem is 146 plus 155. We are adding one hundred, four tens, and six ones to one hundred, five tens, and five ones.
	Examine the ones	In the ones place, there are six ones and I need to decide whether I need to regroup. There are more than ten, so I must trade ten ones for one ten. I mark this in the tens place.
	Note the ones	After regrouping, I have 1 one, so I write that in the ones place.
	Address the tens	In the tens place, I have four tens and I add five tens, and the ten that I regrouped earlier. Four, five, and one is 10 tens. Since I have ten, this is one hundred and I need to regroup. I mark this in the hundreds place.
	Mark the tens	I mark the tens, but there are none after trading ten tens for one hundred. I write zero in the tens place.
	Examine the hundreds; exit with quick check	In the hundreds place, I add one hundred, one hundred, and the one hundred that I regrouped. The answer is three, so I write that in the hundreds place. I must check my answer. My answer should be bigger than the original number since the operation was addition. 301 is bigger than 146. I had almost 150 and I added a bit more than 150 and my answer is about 300. My answer is 14 and that makes sense.

Figure 4–13. *Application of the RENAME Strategy in Solving an Addition Problem at the Abstract Level.*

Problem: The second graders were in charge of bringing cookies to the school picnic. On Thursday, they brought 157 cookies. On Friday, students brought 114 more cookies. How many cookies did the second graders bring for the picnic?

Think Aloud: *The problem asks about how many cookies the second graders brought. There is information that cookies were brought on Thursday and Friday. Some cookies were brought on Thursday and more were brought on Friday. Groups were joined, and we do not know the total. The operation is addition because we are joining groups together to find the total.*

Concrete Level Addition Instruction Using Alternate Numbers
Present problem and represent using base-ten blocks and a place value mat that has been divided into two parts (one for each number). 157 + 114

Hundreds	Tens	Ones
[flat]	[5 rods]	[7 cubes]
[flat]	[1 rod]	[4 cubes]

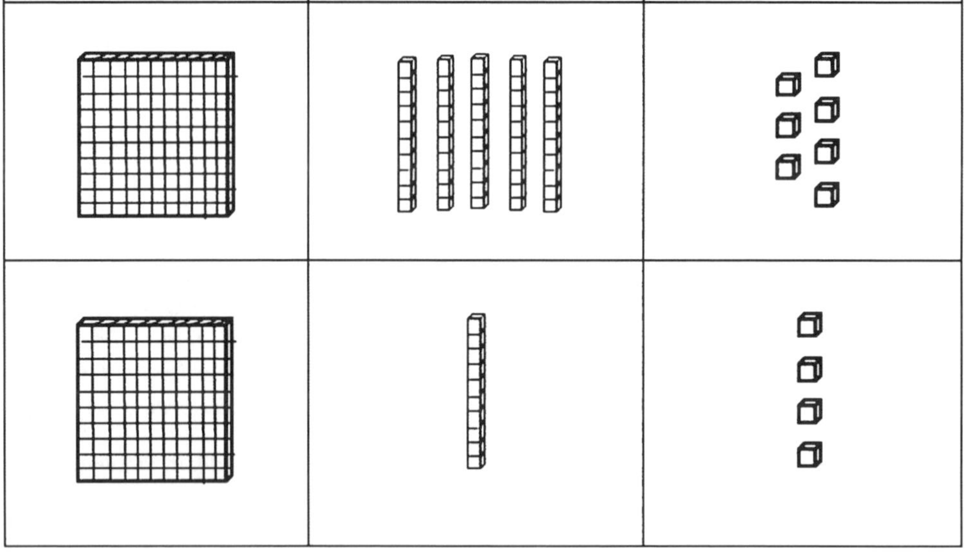

Figure 4–14. Addition With Regrouping Instruction Using Alternative Numbers at Concrete and Representational Levels. continues

Manipulate the base-ten blocks in order to make an alternative problem. Transform the number 157 into 160 by moving three ones blocks from 114 and add them to 157. Ten ones are exchanged to form a ten.		
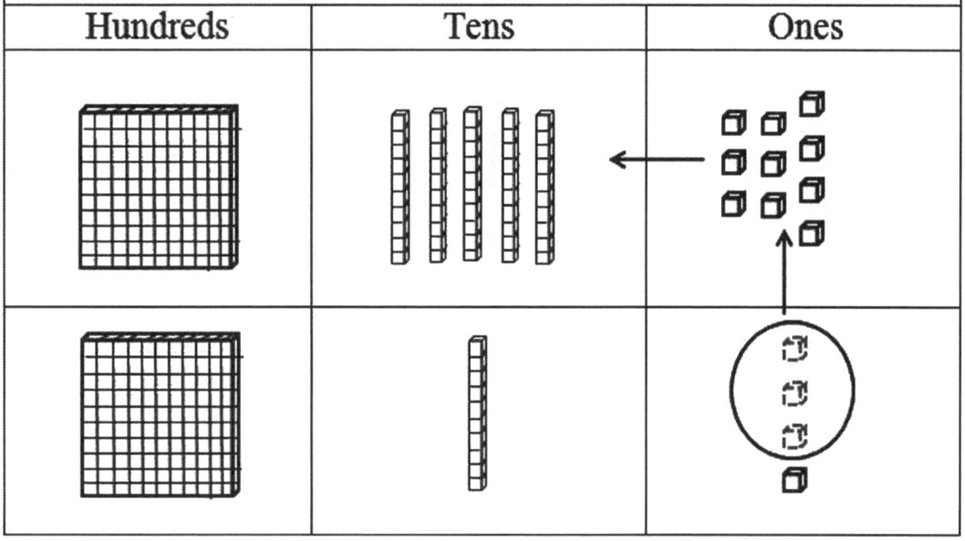 Hundreds	Tens	Ones

The three blocks that were moved transform 157 to 160. The new problem is 160 + 111		
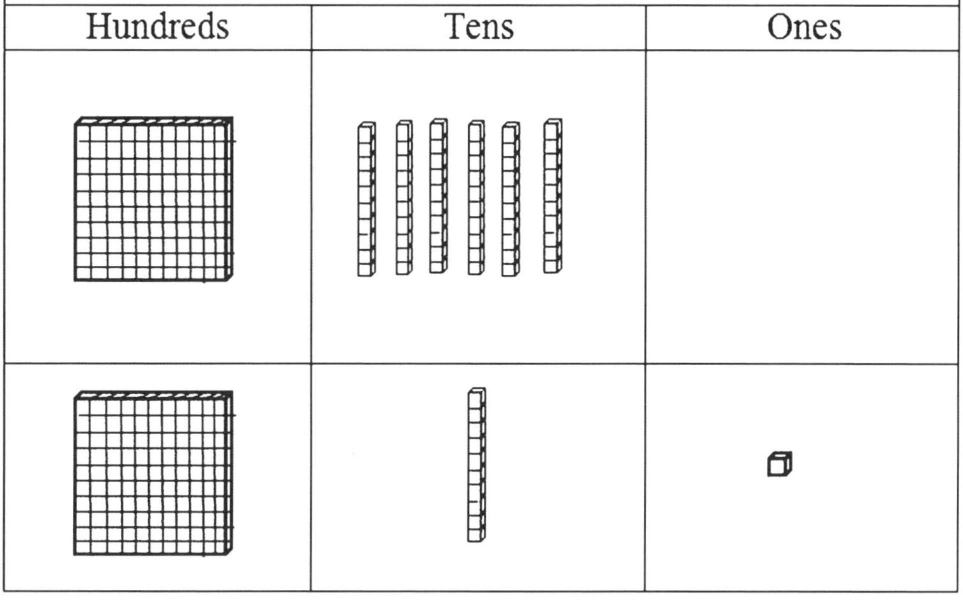 Hundreds	Tens	Ones

Figure 4–14. continues

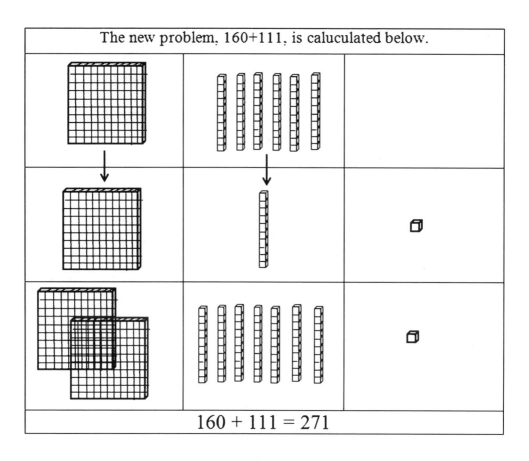

Figure 4–14. continues

Representational Level Addition Instruction Using Alternative Numbers

Manipulate the drawings in order to make an alternative problem. Transform the number 157 into 160 by moving three tallies from 114 and add them to 157. Ten ones are exchanged to form a ten.

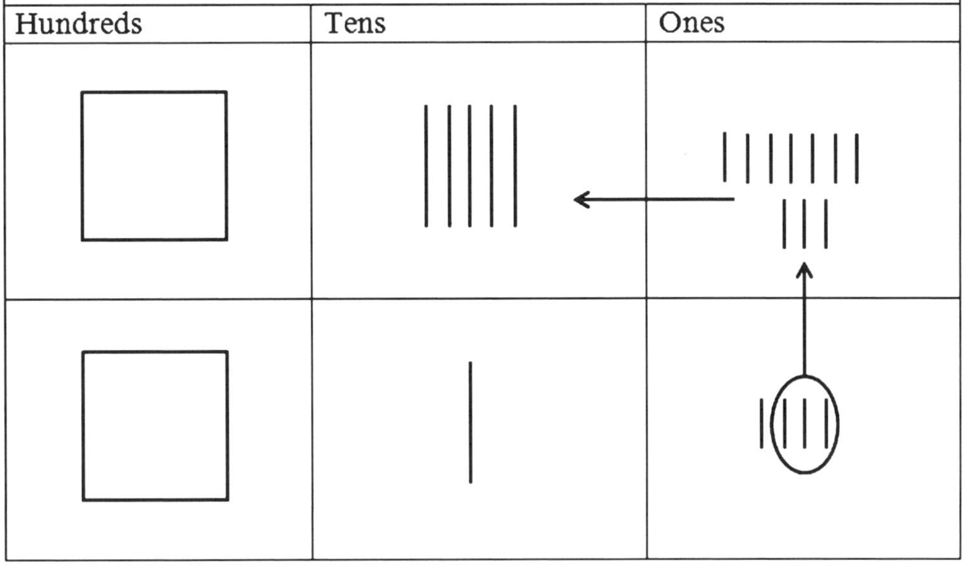

Hundreds	Tens	Ones
☐	||||	||||||| ← |||
☐	|	⦗|||⦘ ↑

The three tallies that were moved transform 157 to 160. The new problem is 160 + 111

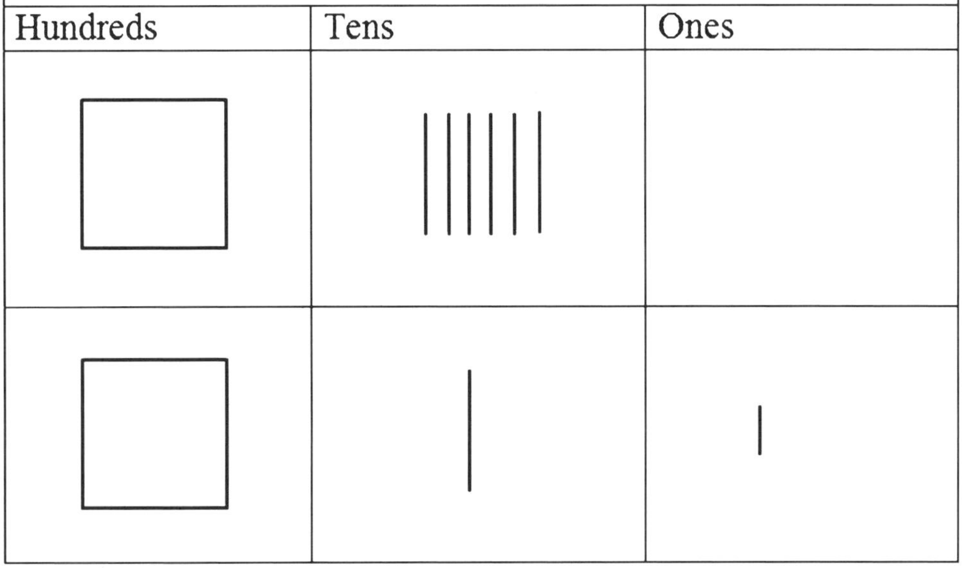

Hundreds	Tens	Ones
☐	||||||	
☐	|	|

Figure 4–14. continues

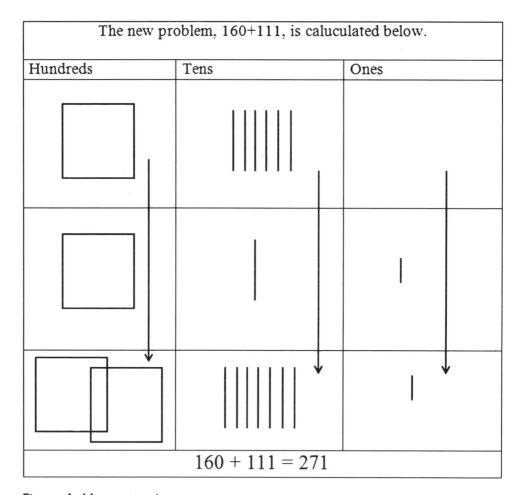

Figure 4–14. continued

PROGRESS MONITORING

Assessing student progress in addition is critical to later success in operations and other mathematics processes. In order to ensure that students understand the operation conceptually, progress monitoring should include formative assessment and observation, rather than just tracking correct answers. For example, students may memorize their basic facts but lack conceptual understanding that provides the foundation for more complex operations as well as application of addition in authentic situations. Observing students as they solve problems at the various levels of the CRA/CSA sequence will inform decisions about the need for intervention and remediation.

Single-Digit Addition

Readiness for addition should be assessed by observing the student's understanding of numbers and counting. First, the student should identify numbers and count in order; these are rote skills. Assess this skill by asking the student to count; the student should say numbers in order. Upon presentation of number symbols, the student should accurately identify each. Conceptual understanding of numbers involves knowledge of quantity, and assessment involves presentation of a particular amount, asking the student to count. The student should demonstrate the ability to touch each object and count with each touch. The student should be able to compare different amounts of objects, telling which is more or less and then compare numbers when presented with number symbols without physical objects present.

Firm understanding of numbers is required before addition is introduced; it is important that these skills extend beyond rote counting. Instruction using the CRA/CSA sequence is introduced, and assessment of understanding begins at the concrete level. Watch and listen as the student reads the problem, translating the mathematical symbols into words. Observe the student physically combine objects. The student should accurately describe his or her actions, combining objects. When asked about the symbols within the addition equation, the student should describe the meaning of the + sign (tells that objects are joined or combined) and = sign (tells that amounts on either side are the same). Description of one's actions while solving problems at the concrete and representational/semi-concrete level will provide information about the student's conceptual understanding. Mastery of addition required for more complex mathematics skills involves both conceptual and declarative knowledge; the student should solve single-digit problems quickly and accurately. The student should identify an answer within 2 seconds of being presented with a problem. Assessment of fluency involves presentation problems with a time limit; give the student a sheet with 30 or more problems and ask the student to solve as many as possible within 1 minute.

The student should also identify the need for addition within real-life situations. When given the following situations, the student should use the correct operation: (a) amounts are compared and the larger amount is missing, (b) there is a whole or class and parts or members of the class are given and the whole or class is missing, and (c) there is a change in which amounts are joined. The student's identification of addition as the operation should not include description of key words (e.g., The words *in all* or *total* appear in the problem; therefore, the operation is addition). The use of key words to identify operations represents a lack of conceptual understanding and instead tells that the student has memorized certain key features for some word problems rather than internalizing the meaning of the operation.

Provide the student with an addition scenario verbally or in writing and ask the student to solve the problem aloud and ask probing questions throughout the process such as the following: (a) What is happening? (b) What are you solving for? (c) What does that number mean? (d) Are there more or less? (e) Who has the most or who has the least? Instruction in addition should begin with translation of words into mathematical symbols, and throughout this beginning instruction, ask students what is happening within the scenario or problem. Doing so will allow for a more seamless development of computation and problem-solving skills.

Addition With Regrouping

As students add larger numbers, understanding of numbers related to the base 10 system is required. Regardless of the approach used to complete the addition process, the student must demonstrate understanding of the composition of numbers. The student must understand that the number, 156, is one hundred, five tens, and six ones rather than the three numerals one, five, and six. The student must also understand how numbers change during addition when there are 10 or more. For example, 13 tens are one hundred and three tens. Assess this knowledge at the concrete and representational/semi-concrete levels of instruction by asking the student to talk aloud while completing a problem. Ask probing questions about the composition of numbers and how the numbers will change throughout the addition process. With the physical and visual aids provided during concrete instruction, the student can more easily use words to describe his or her actions. Therefore, complete a similar observational interview during abstract-level instruction.

CHAPTER SUMMARY

The purpose of CRA/CSA instruction is to provide students with conceptual understanding of mathematical operations, which is accomplished through solving problems using objects at the concrete level and drawings at the representational/semi-concrete level. Students who struggle in mathematics often need more practice, discussion support, and more explicit instruction to master concepts related to numbers and operations, and CRA/CSA provides these. This chapter demonstrated how multiple approaches to addition instruction can be made more explicit using the CRA/CSA sequence.

As students' understanding of addition develops, they move from addition of single-digit numbers to addition of large number that require regrouping. As they move through this learning trajectory, students move toward greater emphasis on procedural knowledge. This procedural knowledge cannot be taught effectively until students have conceptual understanding of the addition operation, number sense, and the role of place value in the joining of numbers greater than 10. Research has shown that the combination of CRA/CSA with the mnemonic strategies is effective in leading students who struggle with computation toward efficient problem solving (Miller & Kaffar, 2011). The strategies combined with CRA/CSA within this chapter, DRAW and RENAME, provide students with steps that help them execute mathematical procedures associated with traditional algorithms. These strategies will only be helpful to students if they have developed conceptual understanding of the operation that is accomplished within the concrete and representational phases of instruction. The authors cannot put enough emphasis on the importance of firm conceptual understanding prior to the introduction of procedural strategies. Providing students who struggle with just a procedural strategy will do more harm than good. Before introducing a strategy, ensure that students can use concrete or representational models to solve problems and explain their mathematical thinking and actions to solve a problem.

Finally, instruction for students who struggle with mathematics must include application of the addition operation within real-life contexts. Word problems must be included in all instruction from basic addition through that of complex addition when regrouping is required. The inclusion of word problems cannot be a set of problems provided as an afterthought at the end of lessons. As mentioned at the beginning of this chapter, they should supersede the presentation of simple equations. A variety of problem types should be explored through word problems as students move along the addition learning trajectory. They should explore problems that reinforce the equal sign as a balance, rather than the sign before the answer. The language of mathematics and its translation to everyday communication should be infused throughout instruction since students who struggle may not generalize these skills without instruction and repeated practice and exposure. In summary, addition instruction should emphasize conceptual understanding, application within authentic situations, and procedural knowledge. Using the CRA/CSA sequence regardless of the approach, traditional or otherwise, will deepen your students' understanding of mathematics and provide them with the skills to compute addition problems but, more important, communicate their knowledge of the operation and apply their knowledge to authentic, everyday situations.

APPLICATION QUESTIONS

1. How are number sense and understanding of place value related to conceptual understanding of addition?

2. Using any approach, how would instruction in beginning addition be taught at each of the levels of the CRA/CSA sequence?

3. What is the rationale for using place value tables or mats when teaching addition with regrouping; how do these tools benefit students?

4. Provide reasons why one might choose each of the following approaches to addition with regrouping: traditional algorithm, decomposition of numbers, or moving numbers.

5. How would a teacher think aloud and provide a rationale for choosing to solve a word problem using addition?

REFERENCES

Barlow, A. T., & Harmon, S. (2012). CCSSM: Teaching in grades 3 and 4. *Teaching Children Mathematics, 18,* 498–507.

Common Core State Standards Initiative (CCSSI). (2010). *Common Core State Standards for Mathematics.* Washington, DC: National Governors Association Center for Best Practices and the Council of Chief State School Officers. Retrieved from http://www.corestandards.org/assets/CCSSI_Math%20Standards.pdf

Dacey, L., & Drew, P. (2012). Common core state standards for mathematics: The big picture. *Teaching Children Mathematics, 18,* 378–383.

Hosp, J. L., Hosp, M. K., Howell, K. W., & Allison, R. (2014). *The abcs of curriculum-based assessment: A practical guide to effective decision making.* New York, NY: Guilford Press.

Mann, R. (2004). Balancing act: The truth behind the equals sign. *Teaching Children Mathematics, 11*(2), 65–69.

Mercer, C. D., & Miller, S. P. (1992). Teaching students with learning problems in math to acquire, understand, and apply basic math facts. *Remedial and Special Education, 13*(3), 19–35.

Miller, S. P., & Kaffar, B. J. (2011). Developing addition with regrouping competence among second grade students with mathematics difficulties. *Investigations in Mathematics Learning, 4*(1), 24–49.

Miller, S. P., Kaffar, B. J., Mercer, C. D. (2011). *Strategic math series: Addition with regrouping.* Lawrence, KS: Edge Enterprises.

Miller, S. P., & Mercer, C. D. (1993). Using data to learn about concrete-semi-concrete–abstract instruction for students with math disabilities. *Learning Disabilities Research & Practice, 8*(2), 89–96.

National Mathematics Advisory Panel. (2008). *Foundation for success: The final report of the National Mathematics Advisory Panel.* Washington, DC: U.S. Department of Education.

Witzel, B. S., Ferguson, C. J., & Mink, D. V. (2012). Number sense: Strategies for helping preschool through grade 3 children develop math skills. *Young Children, 67*(3), 89–94.

CHAPTER 5

Teaching Subtraction Using the Concrete-Representational/ Semi-Concrete–Abstract Sequence

OVERVIEW

The concrete-representational/semi-concrete–abstract (CRA/CSA) sequence has been successful in teaching subtraction to students who struggle in mathematics (Flores, 2009, 2010; Flores, Hinton, & Strozier, 2014; Mancl, Miller, & Kennedy, 2012; Mercer & Miller, 1992a; Miller & Mercer, 1993). The success of this intervention may be its emphasis on conceptual understanding through modeling and representation of numbers and operations. Students who struggle in mathematics need increased intensity of instruction and practice. Explicit instruction using CRA/CSA provides students with repeated practice and increased access to teacher direction and guidance in modeling and representing numbers and operations. It is critical that remedial and intervention practices focus on conceptual understanding and meaning of numbers and operations before teaching mathematical procedures (National Mathematics Advisory Panel, 2008). This chapter provides readers with a rationale for using the CRA/CSA sequence in their intervention practices as well as examples to assist with implementation. It describes and shows how the CRA/CSA sequence can be used to teach the traditional subtraction algorithm as well as other approaches that lead to flexible mathematical thinking.

SEQUENCE OF SUBTRACTION INSTRUCTION WITHIN MATHEMATICS STANDARDS

Instruction in subtraction begins in kindergarten and continues throughout the elementary grades as students complete operations with larger whole numbers and eventually fractions and decimals. Early instruction focuses on the concept of separating or taking apart. Next, concrete models show how the process of separating numbers is expressed using symbolic language with equations. Once students can complete simple subtraction equations using numbers, they explore other ways to solve problems; "taking away" is not the only approach. The standards include student use of strategies. In doing so, students explore ways to use their knowledge of addition and the composition of numbers to solve problems. Subtraction becomes more complex as the numbers within equations increase in magnitude and students' knowledge of the base 10 system deepens. At each level of complexity within the subtraction standards, students begin using models and representations, meaning objects and pictures are used. Only after conceptual understanding is mastered do the standards ask that students fluently compute equations using just numbers and procedures such as the traditional algorithm. The standards (CCSSI, 2010) related to subtraction are located in Table 5–1.

DESCRIPTION OF SUBTRACTION AND PREREQUISITE SKILLS

Subtraction is the inverse of addition; it is separation or the dismantling of numbers. As discussed in the previous chapter, students enter school having had experience with the concept of subtraction without formal instruction or knowledge of the associated mathematical symbols. For example, money is spent, snacks are eaten, toys are lost, and candy is dropped. These are all situations in which children have experienced amounts being changed because a certain amount has been separated from an original amount. Within mathematics, this phenomenon is formally communicated using numerals and symbols that represent the operation of subtraction.

The prerequisite skills for subtraction go hand in hand with those of addition such as counting, number sense, and awareness of numerals representing particular amounts. As these concepts are taught, students should have experiences that build their awareness of numbers in ways that will deepen their understanding of both addition and subtraction (Witzel, Ferguson, & Mink, 2012). For example, when learning the order of numbers, students should count forward and backward. Students' awareness of quantity should include those that are more as well as those that are less. When making numbers, as shown in Chapter 3, provide examples in which a number is made by taking away from a larger number rather than just combining numbers (e.g., four is one less than five). The concepts of addition and subtraction are best taught in a synchronous manner so that students can observe their inverse nature. After learning that addition is the combination or joining of objects, teach subtraction as the separation of a group of objects. Addition instruction at the concrete level should be followed by subtraction instruction at the concrete level. Next, addition at the representational/semi-concrete level would be followed by subtraction at the representational/semi-concrete level. Instruction in addition at the abstract level would be

Table 5–1. Subtraction Standards

Understand subtraction as taking apart and taking from.
Represent subtraction with objects, fingers, mental images, drawings, sounds (e.g., claps), acting-out situations, verbal explanations, expressions, or equations.
Solve subtraction word problems, and subtract within 10 (e.g., by using objects or drawings to represent the problem).
Fluently subtract within 5.
Given a two-digit number, mentally find 10 more or 10 less than the number, without having to count; explain the reasoning used.
Subtract multiples of 10 in the range 10 to 90 from multiples of 10 in the range 10 to 90 (positive or zero differences), using concrete models or drawings and strategies based on place value, properties of operations, and/or the relationship between addition and subtraction; relate the strategy to a written method and explain the reasoning used.
Fluently subtract within 100 using strategies based on place value, properties of operations, and/or the relationship between addition and subtraction.
Subtract within 1,000, using concrete models or drawings and strategies based on place value, properties of operations, and/or the relationship between addition and subtraction; relate the strategy to a written method. Understand that in adding or subtracting three-digit numbers, one adds or subtracts hundreds and hundreds, tens and tens, and ones and ones, and sometimes it is necessary to compose or decompose tens or hundreds.
Mentally subtract 10 or 100 from a given number 100 to 900.
Explain why subtraction strategies work, using place value and the properties of operations.
Use place value understanding and properties of operations to perform multidigit arithmetic.
Fluently subtract within 1,000 using strategies and algorithms based on place value, properties of operations, and/or the relationship between addition and subtraction.
Fluently subtract multidigit whole numbers using the standard algorithm.

followed by subtraction at the abstract level. Fluency in addition facts assists students in becoming fluent in their subtraction facts. Students learn about fact families in which all numbers in the family are related by addition and subtraction (e.g., 2, 3, and 5 are a family in which $2 + 3 = 5$, $3 + 2 = 5$, $5 - 2 =$

3, and 5 − 3 = 2). Teaching these two operations together lays the foundation for more complex understanding of how these inverse operations are used to manipulate numbers at later levels of mathematical understanding.

After students develop mastery of subtraction using single-digit numbers, their understanding of subtraction is dependent upon their conceptualization of place value and the base 10 system. Students must know that our number system is organized by groups of 10. Students' conceptualization of this system allows for subtraction of numbers that are larger than nine and when regrouping. When students do not conceptually understand the workings of our base 10 number system, the process of regrouping within subtraction becomes a series of memorized procedures. Students who struggle in mathematics are likely to memorize place value labels without deep understanding of the base 10 system, meaning that they can identify ones, tens, and hundreds when given a three-digit number. However, this type of declarative knowledge does not lead to accurate approaches to solving subtraction problems involving large numbers and regrouping. For example, the subtraction of 27 from 113 involves taking two tens and 7 ones from 11 tens and 3 ones or 10 tens and 13 ones. In completing the traditional subtraction algorithm, a student must recognize that the number 113 can be broken down into its component parts in order to complete the operation. The following student errors in Figure 5–1 show how students' poor conceptual understanding of our number system leads to subtraction errors.

Current mathematics standards do not require that students reach the answer to 113 − 27 in one particular way. One approach is to count up from 27 until reaching 113, keeping track of the numbers added. Another approach is to transform the problem into an easier one and adjusting the answer according to the change (113 − 27 = 86 is made into 110 − 30 = 80 + 3 + 3). These and the traditional approach require that the students understand the base 10 system and the relations between the numbers within it. The following sections will demonstrate how CRA/CSA provides the conceptual understanding of subtraction as well as the procedural knowledge needed to reach fluency in the operation of subtraction.

CRA/CSA APPLICATION IN DEVELOPING UNDERSTANDING OF BASIC SUBTRACTION

Mercer and Miller (1992a) conducted the earliest CRA/CSA research related to basic operations, including subtraction. The CRA/CSA sequence was used to teach subtraction to students who struggled with mathematics by using objects, drawings, and just numbers with the assistance of a procedural strategy (Mercer & Miller, 1992b; Miller & Mercer, 1993). The curricular programs for teaching subtraction emphasized a traditional approach to the operation; however, the application of CRA/CSA is relevant to other subtraction approaches. Mercer and Miller's CRA/CSA subtraction research involved

$\quad\;\; 2\;\;4\;\;1$ $-\;1\;\;5\;\;3$ $\;\overline{\quad\;\; 1\;\;1\;\;2}$	$\quad\;\; 1\;\;{}^{13}\!\!\not{3}\;\;\not{8}$ $-\quad\;\;\;\; 8\;\;6$ $\;\overline{\quad\;\; 1\;\;5\;\;2}$
The student solving the problem above fails to demonstrate understanding of the operation as well as the number system. Subtracting 3 from 1 will not work, so the student subtracts 1 from 3 instead. Since 4 – 5 does not work, the student computes 5 – 4 instead. This student does not regard the numerals as representing specific amounts but simply as symbols that can be manipulated in any way.	This student has some sense of procedures associated with the traditional algorithm, but a lack of number sense is the root of this error. The student begins subtraction in the hundreds place. The student observes that regrouping is needed in the tens place (eight tens cannot be subtracted from three tens); however, the student trades one and transforms 3 tens into 13 tens rather than 31 (three tens and one). According to this student's procedures, eight tens cannot be subtracted from three tens and a one (31 – 80).

Figure 5–1. Examples of Student Errors Based on Poor Conceptual Understanding of Numbers.

separating one group of objects from another (concrete level) or separating a set of drawings from another (representational/semi-concrete), leaving a difference. As objects (concrete level) and drawings (representational/semi-concrete level) are moved or crossed out, these actions provide students with a visual illustration of the subtraction operation; the language associated with the mathematical operation can be made clear. This provides the student with a mental schema for subtraction, the separation of objects from a group. This schema is first developed with manipulation of objects, and then students draw their own pictures of this process at the representational/semi-concrete level. When students reach the abstract level in which computation involves numbers and symbols only, they have developed conceptual understanding of this abstract concept through their previous hands-on experiences. Basic subtraction instruction should include infusion of real-life situations throughout each level of instruction. Furthermore, students should be taught to discriminate between situations in which objects are joined (addition) and those in which objects are separated (subtraction). When teaching basic subtraction, simple word problems should be translated into equations, rather than only providing printed equations for students to solve. Since subtraction instruction follows conceptual understanding of addition, these simple problems should illustrate situations in which addition is the needed operation and situations in which subtraction is the needed operation. The examples that follow will include these types of problems that will be used to provide in-struction at the concrete and representational/semi-concrete levels. Students who struggle with mathematics have great difficulty generalizing abstract concepts to real-life situations and generally need repeated practice to master mathematical concepts. Therefore, providing students with computation instruction within the context of simple problem solving will provide repeated practice in computation and true problem solving in which the students translate language into mathematical symbols and determine the needed operation. Instruction should provide students with examples of subtraction problems that represent varied situations in which numbers are separated rather than just representing subtraction as "take away." Subtraction should be presented as (a) change (objects are lost, given away, consumed, etc.), (b) part-part-whole (given a class such as students, a subclass such as boys, and missing another subclass such as the number of girls), and (c) comparison in which amounts are measured against each other (more than, less than, etc.). The problem examples within this chapter represent each of these problem types.

Traditional Instruction

Teaching subtraction in a traditional approach using CRA/CSA involves translating the equation into words; the expression $5 - 2$ is "five minus two." Concrete instruction shows the problem-solving process in which five objects are presented and two of the five are separated or taken away from the group, leaving a smaller amount of three as the difference. At the concrete level, the mathematical symbols for equal are explained as well. The symbol, =, and the horizontal line drawn underneath an equation that is written vertically both mean that the amounts on either side of these symbols are the same. The process for concrete basic subtraction instruction is shown in Figure 5–2.

The steps shown in the previous example are taught using explicit instructional procedures, meaning that a teacher introduces the concept and models and guides prior to students' independent practice. Introduction to the concept involves a short review of prerequisite skills, explanation of the steps that will follow, and expectations for student behavior. This introduction is called an advance organizer, which provides students with structure and predictability and provides the teacher with information about the students' readiness for the lesson. The next step in explicit instructional procedures is modeling and demonstration by the teacher. The teacher physically models

Present and translate the problem.	Representation With Manipulatives	Separate and count the difference to arrive at an answer. Ensure that the amounts on either side of the equal symbol are the same.
Will had four pencils in his backpack. Two pencils fell out of his backpack on the way to school. How many pencils does Will have now? 4 − 2 = Will had four and two are lost. Four minus two.	4 − 2  = ▫ ▫ ▫ ▫	4 − 2 = 2 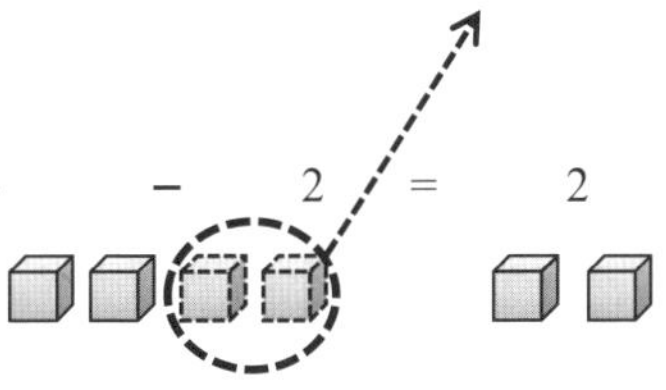 ▫ ▫ (▫ ▫) ▫ ▫

Liza has four papers for homework. She has two more papers than Joe. How many papers does Joe have?
 	
  4
− 2
─────
 	
Liza has four papers. We do not know how many Joe has, but we know that she has two more than Joe, which means he has two less than Liza.
Four minus two.
	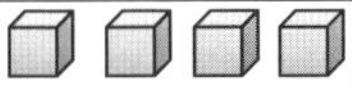 ▫ ▫ ▫ ▫
	  4
	− 2
	─────
		▫ ▫ (▫ ▫)
		  4
		− 2
		─────
		  2 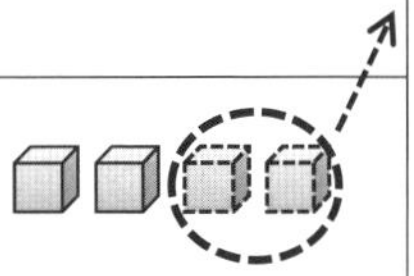 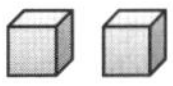
		
		▫ ▫

Figure 5–2. Basic Subtraction Instruction at the Concrete Level.

completion of the problem-solving process and computation as well as demonstrates the mental processes through thinking aloud. Although the teacher models, it is important that the students remain engaged. Student engagement is maintained through repetition of the teacher's vocabulary, counting objects as the teacher manipulates them, or naming the numbers that the teacher writes within a problem. After modeling, the teacher guides students by solving problems with them in a back-and-forth turn-taking process. The teacher provides prompts for students so that they can participate; prompts can be verbal or nonverbal by pointing or using gestures. Next, the teacher presents problems to students for independent practice in which they use objects to solve subtraction problems without assistance. Finally, the teacher provides a wrap-up or postorganizer in which the highlights of the lesson are reviewed.

After mastery of subtraction at the concrete level, which is defined by Mercer and Miller (1992b) as solving problems independently with 80% of answers correct, students are taught to solve subtraction problems using drawings. The students' dependence on physical objects is faded with the representational/semi-concrete level of instruction. When given an equation (based on a simple word problem that is written or presented verbally), the minuend (larger number) is represented using simple drawings such as tallies. The subtrahend is represented by crossing out tallies that correspond with that number. The tallies that remain untouched represent the difference. The process for representational/semi-concrete-level subtraction instruction is shown in Figure 5–3.

Present and translate the problem.	Representation With Manipulatives	Join and count the sum to arrive at an answer. Ensure that the amounts on either side of the equal symbol are the same.
Terrance had red and blue markers. He had six markers in all and two of them were red. How many markers were blue? 6 − 2 = Terrance has two kinds of markers, red and blue. Combined, he has six markers; some are blue and two are red. To find the number of blue markers, we separate the two red ones from the total of six. Six minus two.	6 − 2 = ||||||	6 − 2 = 4 ||||+⊬ ||||
There are usually six students who walk home from school. Today, two of those students were absent. How many students walked home from school today? 6 − 2 ――― The number of students who walk home is six, but two are not part of the whole group because they are absent. We separate 2 from 6 to find the number of students who will walk home today. Six minus two.	6 − 2 |||||| ―――	6 − 2 ||||+⊬ ――― 4 ||||

Figure 5–3. Basic Subtraction Instruction at the Representational Level.

Counting Up

Another approach to subtraction is counting up, and this can be taught using the CRA/CSA sequence as well. In contrast to the traditional approach of separating numbers in which the subtrahend is physically removed from the minuend, counting up involves counting forward from the subtrahend until the amount represented by the minuend is reached. When given the problem, 8 – 3, one counts forward from the subtrahend, three, until reaching eight, the minuend (4, 5, 6, 7, 8). There are five numbers between three and eight, so the answer or difference between them is five. As with the counting approach used in addition, it important that instruction emphasize that counting begin with the number that follows the subtrahend rather than beginning the counting process with the subtrahend (the nonexample provides an answer of 6 given the problem 8 minus 3, because the counting involves 3, 4, 5, 6, 7, 8). This error pattern reliability provides an answer that is one more than the correct answer. Providing instruction in this alternative approach using the CRA/CSA sequence allows students to observe the physical representation of the subtrahend and begin counting with the next object or drawing that is added, decreasing the likelihood that the error pattern will develop. The process for concrete counting-up subtraction instruction is shown in Figure 5–4.

Within representational/semi-concrete instruction, objects are replaced with simple drawings. The subtrahend is represented using tallies; additional tallies are drawn, counting forward until reaching the minuend. The process for representational counting-up addition instruction is shown in Figure 5–5.

Alternative Facts

Another approach to subtraction is using alternative facts to find the difference between two numbers. This approach is especially helpful when one cannot remember a given fact and uses known facts to find the answer. Manipulating numbers, as in the counting-up approach, involves understanding of numbers, their composition, and the base 10 system. For example, when given the problem 18 – 9, a student would use his or her knowledge of numbers and operations to solve a problem using alternative facts such as 18 – 10. Using this approach involves transforming 18 – 9 to 18 – 10 = 8 + 1 = 9. In order to complete this problem accurately, the student must understand that changing the subtrahend requires that the answer be adjusted accordingly. The example problem involved adding one to the subtrahend of 9, which required one to be added to the answer. If a problem is changed by subtracting from the subtrahend (15 – 6 to 15 – 5), the answer is adjusted through subtraction (15 – 6 to 15 – 5 = 10 – 1 = 9).

Problems can also be transformed by changing the minuend, and this change results in changes to the answer. For example, adding to the minuend

Problem Translation and *Think Aloud*

Jay's mom bought 12 doughnuts for breakfast. Jay's brothers and sisters ate 9 doughnuts while he was sleeping. How many doughnuts were left for Jay?

The problem asks about the number doughnuts left for Jay. We know how many doughnuts there were before anyone ate them, and we know that some were eaten before Jay woke up. So there was a big number, some went away, and some are left. We need to separate 9 from the big number, 12, to find the missing number. When we separate numbers, the operation is subtraction. So our problem is 12 − 9. We can also find the missing number by solving this problem: 9 + _?_ = 12.

Concrete Level	
Begin with subtrahend and count up to minuend	12 − 9 = 9 + _?_ = 12
Tens	Ones
	 ▫▫ ▫▫ ▫▫ ▫ 9 + 1 = 10 ▫▫
 ▯	 ▫ ▫ 10 + 2 =
9 + ___ = 12 1 + 2 9 + 3 = 12 or 12 − 9 = 3	

Figure 5–4. Concrete Instruction for Subtraction Using Counting-Up Approach.

(17 − 9 to 20 − 9) results in the need to subtract from the answer (17 − 9 to 20 − 9 = 11 − 3 = 8). In contrast, subtracting from the minuend (12 − 5 to 10 − 5) results in adding to the answer (12 − 5 to 10 − 5 = 5 + 2 = 7). Situations in which this approach is most useful are those in which a problem

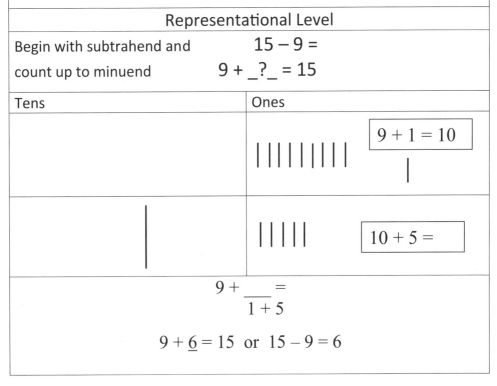

Figure 5–5. Representational Instruction for Subtraction Using Counting-Up Approach.

can be transformed into an easier problem that can likely be solved using mental mathematics. However, before this process becomes automatic and accurately performed mentally, students who struggle in mathematics may need instruction using the CRA/CSA sequence. The process for concrete alternative fact subtraction instruction is shown in Figure 5–6.

Represent the minuend using base 10 blocks.			
14 − 5	Hundreds	Tens ▊	Ones ◻ ◻ ◻ ◻

The current problem involves an unknown fact. By manipulating the numbers, an alternative fact is used to find the answer. If the minuend is changed from 14 to 15, this problem is transformed into an easier subtraction problem.

	Hundreds	Tens	Ones
1 4 +1 = 15 − 5		 ▊	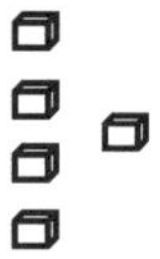 ◻ ◻ ◻ ◻ ◻

Figure 5–6. Basic Subtraction Instruction Using Alternative Facts at Concrete Level. continues

CRA/CSA APPLICATION FOR SUBTRACTION WITH REGROUPING

Research demonstrates that instruction using the CRA/CSA sequence to teach subtraction with regrouping is effective for students who struggle in mathematics (Flores, 2009, 2010; Flores et al., 2014). The CRA/CSA sequence supports and bolsters students' understanding of numbers, place value, and their role within subtraction when the operation becomes more complex with large numbers and regrouping. At the concrete level, problems are solved using base 10 blocks, which have been found to be best organized with a place value mat, as shown in Chapter 4 for addition. The CRA/CSA research demonstrating the effectiveness of CRA/CSA and place value mats has involved instruction in the traditional algorithm, which uses various

	Hundreds	Tens	Ones

1 5̸4
  − 5
    −1
		(one ten rod)	(4 ones and 1 one)

Changing the minuend requires an adjustment to the final answer. Change the minuend by adding 1 and reflect this by subtracting that amount from the final answer.

Complete the new problem (15 − 5). Write the answer to the new problem and subtract the adjusted amount to arrive at the final answer.

	Hundreds	Tens	Ones
1 5̸4 − 5 1 0 −1 10 − 1 = 9		 (one ten rod)	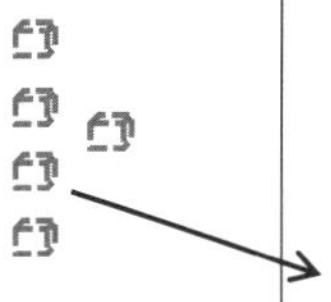 (ones)

Figure 5–6. continued

shortcuts in solving problems (Flores et al., 2014; Mancl et al., 2012). However, the place value mat assists students in organizing base 10 blocks, as well as executing procedures involved in solving subtraction with regrouping problems across multiple approaches, which are shown later in this chapter.

Physically manipulating objects on the place value mat allows students to see changes in numbers, which gives meaning to language that is used during the problem-solving process. The abstract term, "regrouping," can be observed. The student can physically see the need for regrouping when attempting to subtract a larger number from a smaller number because there are too few objects to complete the operation. The student physically removes a block from the column to the left (e.g., ten, hundred, etc.) and exchanges the larger block for the same amount of the smaller blocks (e.g., trading a 10 for 10 ones). While the students physically engage in this process, they can assign words and language to their actions. The mathematical language becomes meaningful and the concept of regrouping has meaning. Students who struggle in mathematics often have great difficulty making meaning of mathematical concepts without sufficient concrete experiences. Without sufficient time and intensity of instruction at the concrete level, students are left to memorize steps and procedures without knowing how or why they are used; difficulties are further compounded when students' memories fail and the procedures are not followed or implemented incorrectly. An example of concrete instruction in subtraction with regrouping using the traditional algorithm is shown in Figure 5–7.

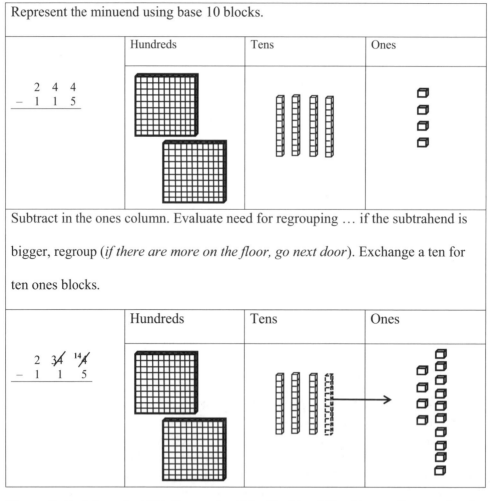

Figure 5–7. *Subtraction With Regrouping Using Traditional Algorithm at Concrete Level.* continues

Complete the operation in the ones column and mark the difference in the written problem.			
	Hundreds	Tens	Ones
2 3̸4 ¹⁴4̸ − 1 1 5 9			

Subtract the tens column. Evaluate the need for regrouping (*there are NOT more on the floor*). Note the difference in the written problem.			
	Hundreds	Tens	Ones
2 3̸4 ¹⁴4̸ − 1 1 5 2 9			

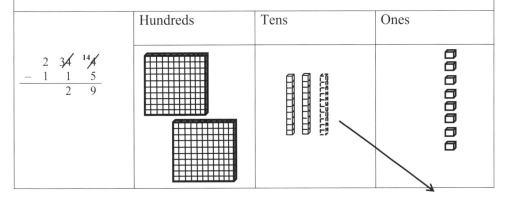

Figure 5–7. continues

Subtract the hundreds column. Evaluate the need for regrouping (*there are NOT more on the floor*). Note the difference in the written problem.				
2 3̸4 ¹⁴4̸ − 1 1 5 —————— 1 2 9	**Hundreds**	**Tens**		**Ones**

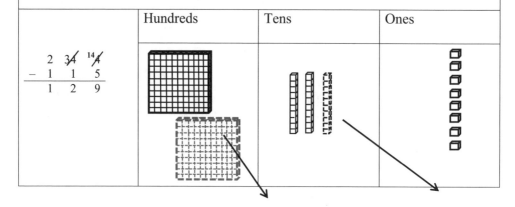

Check the answer. Does the answer make sense? (*Two hundred and about forty minus one hundred and about ten is one hundred and about thirty.*) Does the written answer match the blocks? (*one hundred, two tens, nine ones*)			
2 3̸4 ¹⁴4̸ − 1 1 5 —————— 1 2 9	**Hundreds**	**Tens**	**Ones**

Figure 5–7. continued

Concrete-level instruction focuses on conceptual understanding of the subtraction operation as large numbers are separated and regrouping is required. To continue reinforcement of conceptual understanding, representational/semi-concrete instruction involves drawing problems using a place value chart. However, dependence on objects is faded and the students' role in making meaning of the operation increases because they must draw the representations of numbers. The place value mat is replaced by a chart on which representations are drawn. An example of representational instruction in subtraction with regrouping is shown in Figure 5–8.

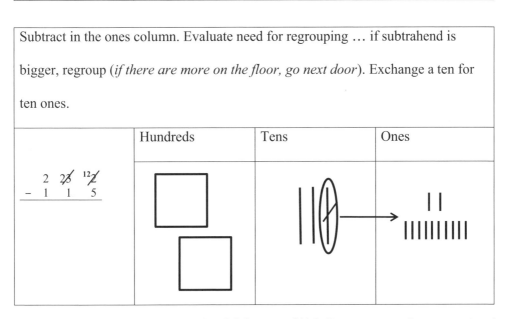

Figure 5–8. Instructional Example of Subtraction With Regrouping at Representational Level. continues

Complete the operation in the ones column and mark the difference in the written problem.			
	Hundreds	Tens	Ones
$\begin{array}{r} 2\ \cancel{2}\ {}^{12}\cancel{7} \\ -\ 1\ \ 1\ \ \ 5 \\ \hline 7 \end{array}$	☐ ☐	∣∣∣∕	╫ ∣∣∣∣∣∣∣╫

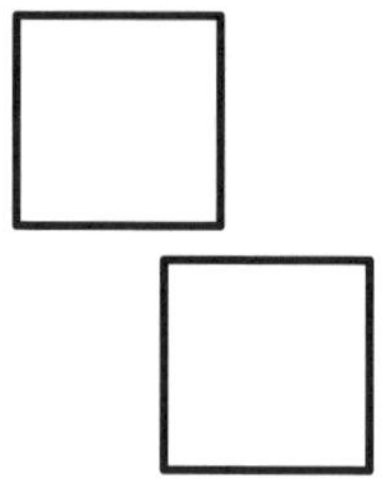

Subtract the tens column. Evaluate the need for regrouping (*there are NOT more on the floor*). Note the difference in the written problem.			
	Hundreds	Tens	Ones
$\begin{array}{r} 2\ \cancel{2}\ {}^{12}\cancel{7} \\ -\ 1\ \ 1\ \ \ 5 \\ \hline 1\ \ \ 7 \end{array}$	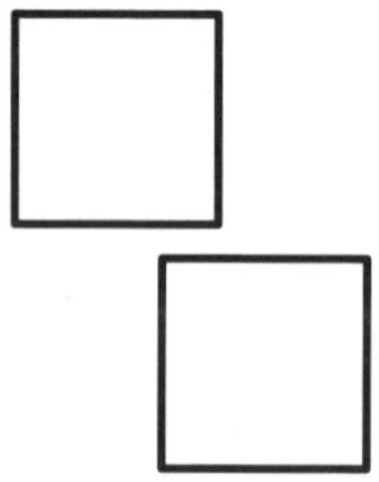 ☐ ☐	 ∣∕∕∕	 ╫ ∣∣∣∣∣∣∣╫

Figure 5–8. continues

Procedural knowledge is required to complete the traditional subtraction algorithm at the abstract level in which problems are solved using just numbers without visual aids. Students must have conceptual knowledge of the operation and the algorithm before procedural knowledge is emphasized. Without conceptual understanding, students who struggle in mathematics will have great difficulty solving problems and developing more complex mathematical skills if only relying on sets of procedures for computation.

Even with conceptual knowledge, students who struggle in mathematics may have difficulty remembering and following procedures related to the traditional algorithm. Miller and Kaffar (2011) developed a mnemonic strategy to be used during the abstract level of instruction. The strategy pro-

5. Teaching Subtraction Using the Concrete-Representational/Semi-Concrete–Abstract Sequence   113

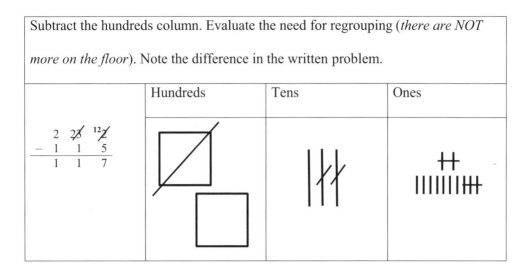

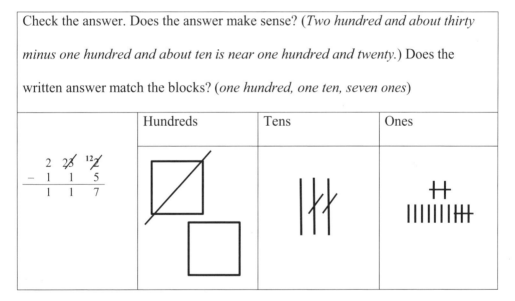

Figure 5–8. continued

vides an efficient set of steps, applicable to any subtraction with regrouping problem. The strategy is taught and used with numbers only, after conceptual understanding has been mastered using base 10 blocks (concrete level) and drawings (representational/semi-concrete). Miller and Kaffar's procedural strategy is as follows: (a) read the problem, (b) examine the ones column and use the *BBB* sentence (if there is a bigger number on the bottom, break a 10 and trade), (c) note ones in the ones column, (d) address the tens column and use the *BBB* sentence (if there is a bigger number on the bottom, break a 10 and trade), (e) mark tens in the tens column, and (f) examine and note hundreds, and exit with a quick check. Figure 5–9 shows application of the RENAME strategy in solving a subtraction with regrouping problem at the abstract level.

The use of the traditional algorithm is one way to approach subtraction with regrouping. Effective mathematics instruction supports the development of flexible thinking (Dacey & Drew, 2012). Therefore, students are taught to use their conceptual knowledge of numbers and operations to solve problems in a variety of ways. The CRA/CSA sequence can support approaches other than the traditional algorithm. When presented with subtraction with regrouping problems, students can solve problems using addition by relying

Problem	Strategy	Think Aloud
0 9 16 1̶ 0̶ 6̶ − 8 8 ――――― 1 8	**R**ead the problem	The problem is 106 minus 88. We are taking 88 from 106.
	Examine the ones	In the ones place, there are six ones and I need to decide whether I can subtract or not. The bottom number is bigger, so I must break a ten and trade. I do not have any tens, so I must break a hundred and trade. Now I have zero hundreds and ten tens. I need to break a ten and trade, so now I have 9 tens and 16 ones.
	Note the ones	After regrouping, I have 16 ones. I subtract eight, so now I have eight left. I write 8 in the ones place.
	Address the tens	In the tens place, I have nine tens and I subtract eight tens. The bottom number is not bigger, so I can subtract. I do not need to regroup.
	Mark the tens	I mark the tens that remain. I write one in the tens place.
	Examine the hundreds; exit with quick check	I have do not have any hundreds. I must check my answer. My answer should be smaller than the original number since the operation was subtraction. Eighteen is smaller than 106. I had a little more than 100 and I subtracted almost 90. I would expect that my answer is more than 10 but less than 20. My answer is 18 and that makes sense.

Figure 5–9. Application of RENAME Strategy.

on their knowledge of operations and numbers. The subtraction process can be bypassed altogether. This counting-up procedure involves beginning with the subtrahend and counting up until arriving at the minuend. The counting-up process was previously shown with regard to basic subtraction but applies to any subtraction situation. When problems involve large numbers, it is helpful to use a place value mat along with base 10 blocks when working at the concrete level. When working at the representational/semi-concrete level, a place value chart assists students in organizing their drawings. Concrete and representational/semi-concrete examples of counting up to solve a subtraction with regrouping problems are shown in Figure 5–10.

Another way to solve subtraction problems is to make an alternative problem. This approach is used to lessen the difficulty of the given problem, but the student must rely on a deep sense of numbers and operations. Changes that are made to a minuend and/or a subtrahend must be accounted for in the answer. Simply changing numbers in problems will not lead to accuracy in computation. Students must recognize the effects of the change in numbers. For example, subtracting 199 from 301 presents a complicated problem when executed following the traditional algorithm and likely requires paper and pencil. An easier problem would be subtracting 200 from 301. However, transforming 199 to 200 requires that the one added to 199 must be subtracted from the final answer. Using the alternative problem approach makes mental mathematics possible; however, learning the concept may require explicit instruction in which the student can see the approach in action. Examples of the alternate problem approach are shown in Figure 5–11.

Problem Translation and *Think Aloud*

Third-grade students went to the book fair. They bought books that were mysteries or adventure stories. Third graders bought 105 books in all. They bought 57 mystery books, and the rest were adventure stories. How many adventure stories did third graders buy?

The problem asks about the number of adventure stories that were bought. We know how many books were bought in all; some were mysteries and some were adventure stories. In all, there were 105 books, and 57 of the books were mysteries. We have the total, but we are missing a smaller part of the total. We need to separate 57 from 105 to find the missing number. When we separate numbers, the operation is subtraction. So our problem would be 105 – 57. We can also find the missing number by solving this problem: 57 + _?_ = 105.

Figure 5–10. Concrete and Representational Examples of Instruction Using Counting Up. continues

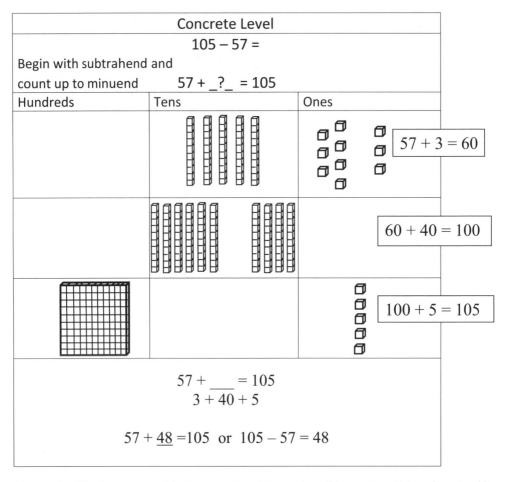

Figure 5–10. *Concrete and Representational Examples of Instruction Using Counting Up.* continues

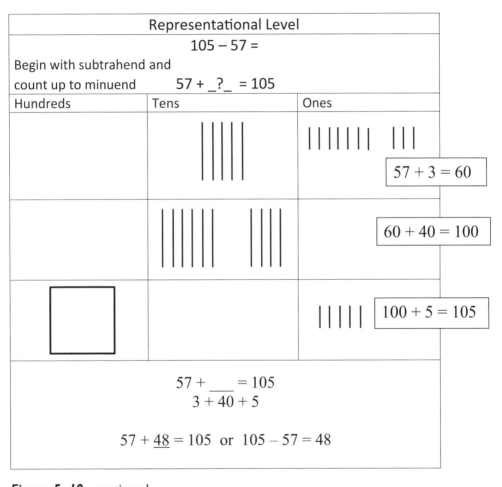

Figure 5–10. continued

Problem and Think Aloud

The students in fourth grade sold 115 tickets to the Fall Festival. The students in the third grade sold 244 tickets to the Fall Festival. How many more tickets did the third graders sell than the fourth graders?

The students in third grade and fourth grade sold tickets. The problem asks us to compare the number of tickets sold in each grade. To find how many more were sold, we must find the difference between the number of tickets sold by third graders and fourth graders. We must separate the numbers. When we separate numbers, the operation is subtraction. So, we subtract 115 tickets from 244 tickets. We write the problem as 244 minus 115.

Represent the minuend using base 10 blocks.			
2 4 4 − 1 1 5	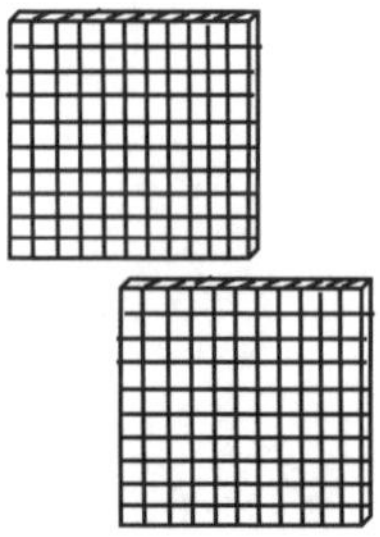 Hundreds	 Tens	 Ones

Figure 5–11. *Concrete and Representational Level Instruction Using Alternative Problem Approach.* continues

		Hundreds	Tens	Ones
The current problem requires regrouping. By manipulating the numbers, regrouping can be avoided. If the minuend is changed from 244 to 245, this problem is transformed into an easier subtraction problem that does not require regrouping.				
2 4 4 − 1 1 5				

		Hundreds	Tens	Ones
Changing the minuend requires an adjustment to the final answer. Change the minuend by adding 1 and reflect this by subtracting that amount from the final answer.				
2 4 5̸4 − 1 1 5 − 1				

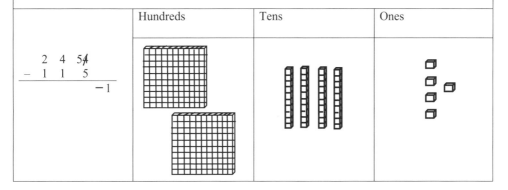

Figure 5–11. continues

Complete the new problem (245 – 115). Write the answer to the new problem and subtract the adjusted amount to arrive at the final answer.			
	Hundreds	Tens	Ones
2 4 5̸4̸ − 1 1 5 ――――― 1 3 0 − 1 130 − 1 = 129	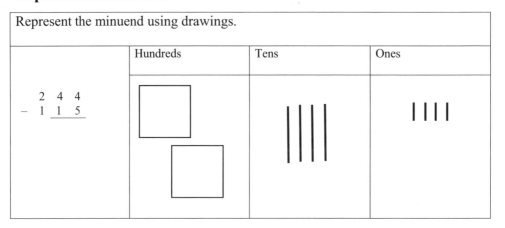		

Representational Level

Represent the minuend using drawings.			
	Hundreds	Tens	Ones
2 4 4 − 1 1 5 ――――	□ □	||||	||||

Figure 5–11. continues

The current problem requires regrouping. By manipulating the numbers, regrouping can be avoided. If the minuend is changed from 244 to 245, this problem is transformed into an easier subtraction problem that does not require regrouping.			
	Hundreds	Tens	Ones
2 4 4 − 1 1 5	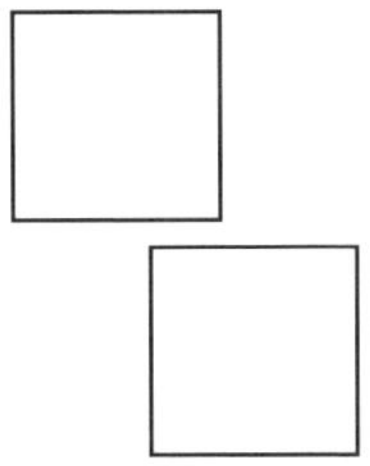 □ □	 ||||	 ||||

Changing the minuend requires an adjustment to the final answer. Change the minuend by adding 1, and reflect this by subtracting that amount from the final answer.			
	Hundreds	Tens	Ones
2 4 5̸4̸ − 1 1 5 − 1	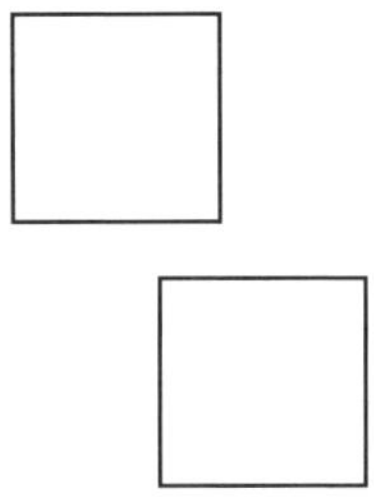 □ □	 ||||	 |||| |

Figure 5–11. continues

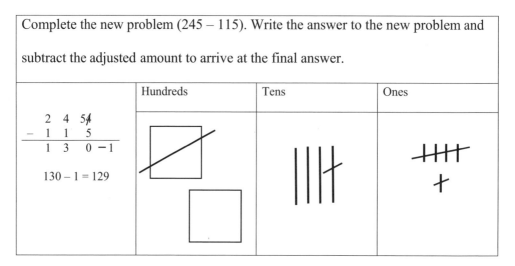

Figure 5–11. continued

PROGRESS MONITORING

Before beginning instruction in subtraction, ensure that students have the prerequisite skills related to counting and number sense. Rote counting and identification of number symbols are fundamental skills, but they are not sufficient for the development of conceptual understanding. The student should demonstrate an understanding of numbers as representing amounts. Furthermore, as the student counts forward, he or she should recognize that the amount is increasing. When counting backward, the student recognizes that the amount decreases. Provide the student with concrete experiences that include counting so that you can observe the student counting accurately (one touch for each count) and ask probing questions about quantity and comparison to another group of objects.

Single-Digit Subtraction

Monitoring the student's progress while completing subtraction should occur throughout the CRA/CSA sequence. During the concrete and representational phases, conceptual understanding of subtraction develops. Ask the student to describe the meaning of symbols as well as his or her actions. Mastery involves description of the subtraction symbol as meaning that an amount will be removed or separated. Ask the student to complete an equation with objects and/or pictures and talk aloud while doing so. Observation of the student separating and counting can be observed. Error patterns can be better identified while watching the student work, asking probing questions. Interpreting completed work samples will provide you with some information but will not provide explanations related to how and why errors occur.

The student should also interpret real-life situations, identifying the need for subtraction. When given the following situations, the student should identify the correct operation: (a) amounts are compared and the smaller amount is missing, (b) there is a whole or class and parts or members of the class are missing, and (c) there is a change in which an amount is separated, or lost. The student's identification of subtraction as the operation should not include description of key words (e.g., The words "less than" appear in the problem; therefore, the operation is subtraction). Dependence on key words demonstrates a clear lack of conceptual understanding of operations. Provide the student with a scenario verbally or in writing and ask the student to solve the problem aloud and ask probing questions throughout the process, such as the following: (a) What is happening? (b) What are you solving for? (c) What does that number mean? (d) Are there more or less? or (e) Who has the most or who has the least? The translation of words into mathematical symbols is difficult and should be included in instruction from the beginning; students should not view computation and solving of word problems as two separate concepts. Rather, equations (5 − 2 =) are the mathematical translations of real-life situations provided through word problems or verbal scenarios.

Subtraction With Regrouping

When monitoring student progress regarding problems that involve regrouping, assess the students' understanding of numbers, the base 10 system, and the need for regrouping. Ask the student to describe a two-digit number as being composed of tens and ones. Observational assessment during concrete- and representational/semi-concrete level instruction will provide information regarding the students' understanding as the student physically manipulates base 10 blocks. Ask the student to tell why regrouping may be necessary (five cannot be subtracted from two because there are not enough). Ask the student to describe how the numbers can be manipulated in order to complete the problem. At the concrete and representational levels, the regrouping process is more easily described given the physical and visual aids that are present. Therefore, the same type of interview should be conducted when given problems at the abstract level. When the student is presented with an equation, ask the student what the numbers represent and how he or she will approach the computation process. The student should identify numbers as being composed of ones, tens, or hundreds; the need for regrouping; and the execution of the regrouping process. The same process should occur when the student uses approaches other than the traditional algorithm such as counting up or use of alternative problems.

CHAPTER SUMMARY

The concrete-representational/semi-concrete–abstract sequence involves solving problems using physical objects and drawings or pictures prior to instruction using just numbers. This allows students to develop conceptual understanding of numbers and operations prior to an emphasis on procedural knowledge and declarative knowledge or fluency and automaticity. Students who struggle in mathematics have not received an adequate amount or intensity of instruction, and the CRA/CSA sequence provides explicit instruction at each level of mathematical understanding (conceptual, procedural, and declarative).

Students first learn about the subtraction operation at a conceptual level, separating groups and working with single-digit numbers. The development of this understanding should occur in concert with their understanding of addition, the joining of groups. The two operations are the inverse of each other. As numbers within equations become larger, regrouping may be required; this involves students' knowledge of our number system in which the base is 10. Conceptual knowledge of the number system and procedural knowledge of how numbers are composed and decomposed are both required for regrouping. It is essential that students have firm conceptual understanding of numbers, the base 10 system, and the operation of subtraction in order to be successful in subtraction of numbers that involve regrouping. This chapter will show how the CRA/CSA sequence provides an avenue in which these concepts can be reinforced for students who struggle in mathematics. Concrete and representational/semi-concrete instruction allows students to observe numbers being separated, the need for regrouping, and the regrouping process. Instruction using CRA/CSA has been combined with strategy instruction to teach procedural knowledge (Flores et al., 2014; Mercer & Miller, 1992a; Miller & Kaffar, 2011). Strategies that assist students in computation of basic subtraction and subtraction with regrouping are provided in this chapter. This chapter also shows how CRA/CSA is used to teach students how to solve subtraction problem in alternative ways, bypassing the regrouping process completely.

Finally, throughout the chapter, each instructional example was shown within the context of application. Students who struggle in mathematics must be shown how operations apply to real-life situations. Instructional lessons must include presentation of computation within the context of word problems or problem situations presented verbally. In order for students to generalize their understanding of the subtraction operation, they must have regular instruction and practice solving problems in which they apply their knowledge of subtraction and discriminate between problems that require subtraction and those that do not require subtraction. Therefore, the examples within this chapter included problem-solving application.

APPLICATION QUESTIONS

1. How are number sense and understanding of place value related to conceptual understanding of subtraction?

2. Using any approach, how would instruction in beginning subtraction be taught at the concrete and representational the levels of the CRA/CSA sequence?

3. What is the rationale for using place value tables or mats when teaching subtraction with regrouping; how do these tools benefit students?

4. Provide reasons why one might choose each of the following approaches to subtraction with regrouping: traditional algorithm, counting up, or alternate problem/numbers.

5. How would a teacher think aloud and provide a rationale for choosing to solve a word problem using subtraction rather than another operation such as addition?

REFERENCES

Common Core State Standards Initiative (CCSS). (2010). *Common Core State Standards for Mathematics*. Washington, DC: National Governors Association Center for Best Practices and the Council of Chief State School Officers. Retrieved from http://www.corestandards.org/assets/CCSSI_Math%20Standards.pdf

Dacey, L., & Drew, P. (2012). Common core state standards for mathematics: The big picture. *Teaching Children Mathematics*, *18*, 378–383.

Flores, M.M. (2009). Teaching subtraction with regrouping to students experiencing difficulty in mathematics. *Preventing School Failure*, *53*(3), 145-152.

Flores, M. M. (2010). Using the concrete-representational-abstract sequence to teach subtraction with regrouping to students at risk for failure. *Remedial & Special Education*, *31*(3), 195–207.

Flores, M. M., Hinton, V. M., & Strozier, S. D. (2014). Teaching subtraction and multiplication with regrouping using the concrete-representational-abstract sequence and strategic instruction model. *Learning Disabilities Research and Practice*, *29*, 75–88.

Mancl, D. B., Miller, S. P., & Kennedy, M. (2012). Using the concrete-representational-abstract sequence with integrated strategy instruction to teach subtraction with regrouping to students with learning disabilities. *Learning Disabilities Research and Practice*, *27*(4), 152–166.

Mercer, C. D., & Miller, S. P. (1992a). Teaching students with learning problems in math to acquire, understand, and apply basic math facts. *Remedial and Special Education*, *13*(3), 19–35.

Mercer, C. D., & Miller, S. P. (1992b). *Strategic math series: Subtraction facts 0–8*. Lawrence, KS: Edge Enterprises.

Miller, S. P., & Kaffar, B. J. (2011). Developing addition with regrouping competence among second grade students with mathematics difficulties. *Investigations in Mathematics Learning, 4*(1), 24–49.

Miller, S. P., & Mercer, C. D. (1993). Using data to learn about concrete-semiconcrete-abstract instruction for students with math disabilities. *Learning Disabilities Research & Practice, 8*(2), 89–96.

National Mathematics Advisory Panel. (2008). *Foundation for success: The final report of the National Mathematics Advisory Panel.* Washington, DC: U.S. Department of Education.

Witzel, B. S., Ferguson, C. J., & Mink, D. V. (2012). Number sense: Strategies for helping preschool through grade 3 children develop math skills. *Young Children, 67*(3), 89–94.

CHAPTER 6

Teaching Multiplication Using the Concrete-Representational/ Semi-Concrete–Abstract Sequence

OVERVIEW

This chapter illustrates how the concrete-representational/semi-concrete–abstract (CRA/CSA) sequence is used to teach multiplication, from basic facts to multidigit problems that require regrouping. Along the learning trajectory, examples of instructional processes at each stage of the sequence, concrete, representational, and abstract, are described and shown. Mathematical thinking, which is necessary to be effective in the processes utilized in multiplication, involves flexibility in choosing among approaches to problem solving (Common Core State Standards Institute [CCSS], 2010; Dacey & Drew, 2012). In order to provide instruction that prepares students to apply the mathematical thinking and practices within each grade level's standards, application of CRA/CSA is shown across different approaches such as area models, partial products, and traditional algorithms. This chapter provides readers with the rationale for choosing the CRA/CSA sequence to teach students who struggle in mathematics as well as the tools to implement instruction.

SEQUENCE OF MULTIPLICATION INSTRUCTION WITHIN MATHEMATICS STANDARDS

Instruction in multiplication begins in second grade, when the foundation for combining equal groups is established. Students manipulate and arrange objects into equal groups and add. For example, students make three rows with two objects in each row and add to find the total. This concept of repeated addition is later taught as a new operation, multiplication, which is the combining of equal groups. In third grade, the mathematical language and symbols associated with multiplication are formally taught usually objects, pictures, and arrays, which provide students with structure for making equal sets. After demonstrating understanding of the concept by making models and representations, students are expected to solve simple multiplication equations quickly and accurately. After mastering multiplication involving single-digit multipliers, students explore multiplication of larger numbers using models that include objects and pictures. Students use their knowledge of the multiplication operation, numbers, and place value to solve problems that include larger numbers and regrouping. The last of the multiplication computation standards includes the use of the standard algorithm, which requires shortened procedures. It is important that students are not asked to use or memorize these procedures without a firm understanding of numbers, place value, and the multiplication operation. The standards (CCSSI, 2010) related to multiplication are located in Table 6–1.

DESCRIPTION OF MULTIPLICATION AND PREREQUISITE SKILLS

The introduction of multiplication leads students to a more complex understanding of numbers. Whereas addition and subtraction involve variations of joining and separating numbers, multiplication is the joining of equal groups of numbers or repeated addition. An understanding of multiplication requires conceptual understanding learned well before mathematics standards related to multiplication are presented. Multiplication requires that

Table 6–1. Multiplication Standards

Determine whether a group of objects (up to 20) has an odd or even number of members (e.g., by pairing objects or counting them by 2s); write an equation to express an even number as a sum of two equal addends.
Use addition to find the total number of objects arranged in rectangular arrays with up to five rows and up to five columns; write an equation to express the total as a sum of equal addends.
Multiply one-digit whole numbers by multiples of 10 in the range 10 to 90 (e.g., 9 × 80, 5 × 60), using strategies based on place value and properties of operations.
Multiply a whole number of up to four digits by a one-digit whole number, and multiply two two-digit numbers, using strategies based on place value and the properties of operations. Illustrate and explain the calculation by using equations, rectangular arrays, and/or area models.
Fluently multiply multidigit whole numbers using the standard algorithm.

the student attends to number symbols as representations of quantities of objects or groups of objects rather than a numeral symbol. Multiplication often begins with making equal groups of objects and translating this into mathematical symbols: three groups of four objects is 3 × 4. As students move along the multiplication learning trajectory, they must have more complex understanding of numbers, place value, and the base 10 system. Simply labeling columns within a table is not sufficient conceptualization for solving more complex multiplication problems that involve numbers larger than 10. Instead, students must understand the relations between numbers within the base 10 system. For example, the problem 6 × 43 is six groups of four tens and six groups of three ones. This requires that students understand that the numeral 4 represents four tens or 40, and the numeral 3 represents three ones and that these numbers are combined to make 43. Then, the problem involves making groups of 40 and three, six groups of three and six groups of 40.

Students who struggle with mathematics may have gaps within their previous learning that interfere with their understanding of multiplication, especially more complex problems in which multiplicands (the number being multiplied) are composed of two-digit numbers, and/or multipliers (the number you are multiplying by) are composed of two-digit numbers. The multiplicand and multiplier are often called factors, without specifying which is which. The CRA/CSA sequence provides scaffolding for students, using physical tools and visual aids that show number concepts at work within the multiplication operation. This chapter shows intervention strategies that address beginning understanding of the operation as well as strategies that address regrouping. The use of the CRA/CSA sequence can be applied to multiple approaches to solving multiplication problems that include partitioning strategies and the standard/traditional algorithm.

Understanding multiplication is more than simply arriving at the correct answer. For example, students may memorize basic facts, making an association between three numbers (e.g., 2 times 3 is 6). Mathematics standards call for explanation of one's computation, which requires that the student can articulate how an answer was found. This chapter includes think-aloud sections that show how the problem-solving process would be explained. This is the outcome for which a teacher would strive. Students who struggle with mathematics will likely be unable to engage in this process without instruction that includes teacher modeling and guidance. This should begin with the teacher modeling the thinking process before asking the student to think aloud independently. The use of the CRA/CSA sequence will assist the student in generating language associated with computation because the student will physically manipulate numbers and have a visual representation of the numbers that are changing throughout the multiplication process. It is critical that readers keep in mind that the students for whom this chapter is written have already received instruction in multiplication but have not been successful; therefore, modeling, guidance, and repeated practice are necessary in order to increase the intensity of instruction.

CRA/CSA APPLICATION FOR BASIC MULTIPLICATION

The early CRA/CSA research focused on basic operations, including multiplication (Mercer & Miller, 1992a; Miller & Mercer, 1993). Within this research, the CRA/CSA sequence was used to model the multiplication operation using manipulative objects, drawings, and numbers only to develop conceptual understanding, procedural knowledge, and eventual fluency in basic multiplication facts. Although the research and the program developed based on the research involve a traditional approach to the operation, making groups of objects, the CRA/CSA sequence can be used to supplement other approaches to teaching the basic concepts of the operation.

Teaching Multiplication by Grouping Objects Using Arrays

Instruction begins with a review and demonstration of repeated addition, adding the same number multiple times (e.g., 2 + 2 + 2 + 2). Within real situations within the classroom, bring students' attention to the use of repeated addition such as when each member of a group needs two pieces of paper or three markers to complete a project. Within the general education setting, multiplication is commonly introduced using the area model. The area model is also sometimes referred to as the array model. The area model allows students to see and connect multiplication to the geometric property of area. The area model uses a grid, and multiplication is depicted with columns as groups and rows as the items within each group; finding the total amount involves filling in the total area of the grid. An example of an area model for the problem, 3 × 4, is shown in Figure 6–1.

Students who struggle with mathematics are likely to have been exposed to this model but are in need of more explicit or more intensive instruction in order to master basic multiplication concepts. Moreover, students will be shown more complex multiplication concepts, division, or fraction concepts using arrays or area models. Therefore, it would be logical that individual or small group interventions using CRA/CSA would include concrete and representational/semi-concrete instruction using arrays rather than just the group-

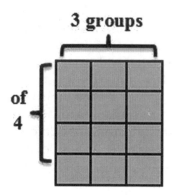

Figure 6–1. Area Model Showing 3 × 4.

ing model shown in the next section. When providing intensive explicit instruction using CRA/CSA and area models, it is important to provide students with language instruction in which mathematical symbols are translated into words in addition to teaching about the conceptual meaning of the operation, the joining of equal amounts. Instruction begins with a review and demonstration of repeated addition, adding the same number multiple times (e.g., 4 + 4 + 4). Multiplication is introduced as a way of showing this repeated addition using numbers and symbols (e.g., 4 + 4 + 4 expressed as 3 × 4). The mathematics expression is translated into words; the expression, 3 × 4, is "three groups of four objects." Concrete-level instruction shows the problem-solving process. Assist students in identifying the groups and objects in the grid. This can be accomplished with prompting questions and is essential to understanding this representation. Using a grid, label the columns across the top as "groups" or the specific amount of groups (three groups) and label the rows on the left side of the grid as the "of objects" (of four). Place one base 10 ones block into each of the four cells within the first column (group). Do the same with the remaining columns, until there are four blocks in each column. The objects are counted to find the product. Concrete multiplication instruction with an array is shown in Figure 6–2.

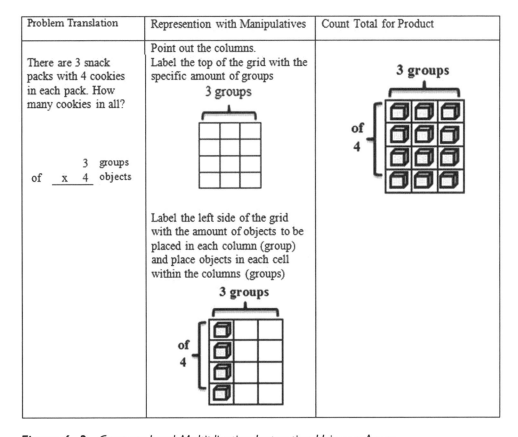

Figure 6–2. Concrete-Level Multiplication Instruction Using an Array.

Representational/semi-concrete instruction would be provided using similar procedures; however, multiplication is represented by shading the array rather than placing objects within the cells. Representational/semi-concrete multiplication instruction with an array is shown in Figure 6–3.

At the abstract level, the focus of instruction would be increasing recall and fluency of multiplication facts. It is important to ensure that students demonstrate an understanding of the concept of joining equal groups prior to abstract instruction. Without this conceptual understanding, students will be unable to solve more complex problems that involve multiplication of larger numbers and regrouping. Students who lack conceptual understanding of the operation (which is taught during concrete and representational/semi-concrete instruction) will be confused with more complex operations, and the student will be left to memorize procedures, an approach that usually results in error patterns and failure.

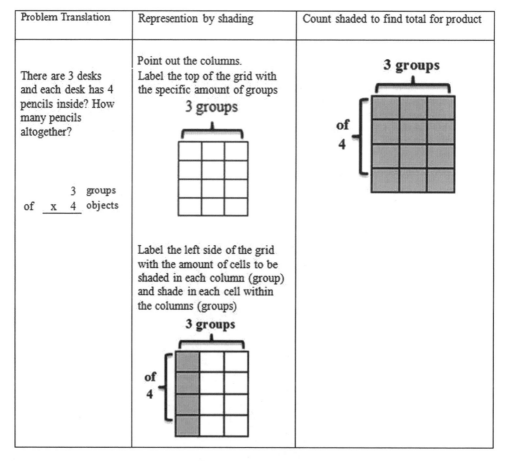

Figure 6–3. Representational-Level Multiplication Instruction Using an Array.

Teaching Multiplication by Grouping Objects

Another approach to teaching the concept of multiplication is by physically grouping objects. This offers another way in which groups of objects can be joined together; instead of using columns and rows, students gather objects and place them in groups. For students who struggle with spatial relations, grouping may be easier.

Multiplication is introduced as a way of showing this repeated addition, as discussed in the previous section. The mathematics expression is translated into words; the expression, 2 × 3 is "two groups of three objects." Concrete-level instruction shows the problem-solving process. Two plates or other physical representations of groups with three objects are placed on each plate. The objects are counted to find the product. The objects used to construct groups could be mathematics manipulative items such as counters or base 10 ones blocks. Students are encouraged to use their mathematical thinking to make sense of the problem. For students who struggle, you will likely guide students in this process, assisting the student in making meaning and using academic mathematical language. Assistance may involve modeling and thinking aloud, completing the task with the student, and prompting the students' language and thinking process. Then students can be successful in responding to questions related to representation of the problem with concrete manipulatives and problem solving. It is important that students attempt and are successful in solving each phase of the problem, including an explanation of their actions and processes. The process for initial concrete multiplication instruction is shown in Figure 6–4.

In order for the concept to be meaningful and relevant to students, descriptions of common situations should accompany written problems. This provides students with the context in which the operation is used and also allows for practice with language and its use in describing operations. If beginning instruction only involves computation problems, students may develop conceptions that computation and solving word problems are different tasks. Using simple contextual situations allows for computation as well

Problem Translation	Representation With Manipulatives	Count Total for Product
2 groups of × 3 objects	 (two circles each containing 3 objects)	 (hand counting two circles each containing 3 objects)

Figure 6–4. Concrete-Level Multiplication Instruction Using Groups.

as instruction and practice with language and its use in solving problems. An example is shown in Figure 6–5.

When students can use manipulative objects to solve multiplication, instruction moves to the representational/semi-concrete level. According to Mercer and Miller (1992b), students need approximately three concrete lessons to reach mastery. Representational-level instruction involves the use of drawings rather than objects. Drawings serve as a visual aid in completing the operation, fading students' dependence on physical objects. The most efficient way of drawing representations of numbers is to use tallies, drawn on a horizontal line. Some intervention materials use more elaborate drawings, but these can interfere with instruction. For example, if using large circles for groups and small circles for objects, students can become consumed or distracted by their drawings by perseverating on the artistic quality, size, shape, or symmetry. In addition, students' poor visual spatial skills or poor motor skills can manifest in drawings that are difficult to interpret, meaning that students will have difficulty counting the objects drawn because they are too

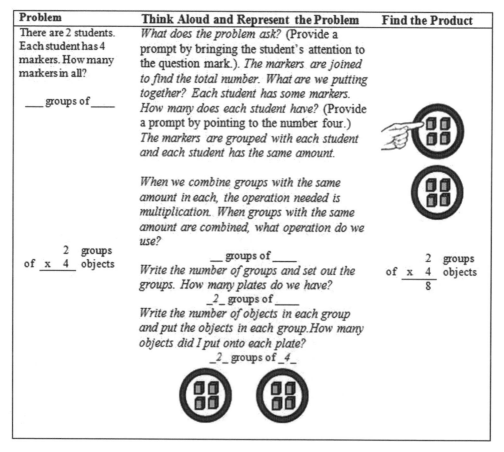

Figure 6–5. Concrete-Level Multiplication Instruction With Contextual Problems.

small, too close together, or too far apart. Therefore, an efficient procedure for drawing multiplication problems is to use horizontal lines to represent groups. Vertical tallies are drawn on each horizontal line to represent the objects in the group. These representations can be drawn simply and in an organized manner, more so than other drawings. However, it is important to ensure students understand that the vertical lines represent the objects in the group and the horizontal lines represent the groups. A student should be able to explain and relate the representation with drawings to concrete examples and real-world examples to demonstrate they understand this representation. An example of a problem solved at the representational/semi-concrete level is shown in Figure 6–6.

After students demonstrate mastery of the operation using drawings, instruction moves to the abstract phase. Curriculum materials used to teach multiplication facts using CRA/CSA define mastery as three lessons in which the student independently solves problems using drawings with at least 80% accuracy (Mercer & Miller, 1992b). However, it is critical that the students are able to answer teacher-initiated questions that focus on meaning, explanation, and creating problems as well. Asking questions in a variety of ways helps to ensure conceptual understanding.

Curriculum materials designed by Mercer and Miller include a mnemonic strategy for solving problems within the abstract phase: (a) discover the sign, (b) read the problem, (c) answer or draw and check, and (d) write the answer (DRAW). The DRAW strategy provides students with a bridge from using visual aids to computation using numbers only. The DRAW strategy also provides students with a set of steps that will increase the likelihood of accurate computation by (a) bringing the student's attention to the symbol and the numbers within the problem, a frequent mistake of students who struggle, and (b) reminding the student of the option to draw the problem if the answer cannot be recalled by memory. During the abstract phase, students solve problems using the DRAW strategy, and the focus of instruction is fluency in basic facts. Since the conceptual foundation of the operation has been laid within the concrete and representational/semi-concrete phases, abstract-level instruction involves teaching students about rules and properties that will assist in memorization of multiplication facts. These include (a) the zero rule (any number multiplied by zero is zero), (b) the one times rule

Problem Translation	Representation With Drawings	Count Total for Product
of x 2 groups 4 objects	|||| ||||	|||| |||| of x 2 groups of 4 objects 8

Figure 6–6. Representational-Level Multiplication Instruction Using Groups.

(any number multiplied by one equals the original number), (c) the doubles rule (any number multiplied by two equals the number added to itself), and (d) the order rule or commutative property, which states that changing the order of numbers within the problem does not change the answer. Teaching these rules can involve providing multiple situations that utilize one of these rules and asking questions to support the student's discovery of the pattern, which is an important mathematical practice. Instruction may also involve drill and practice activities and exercises that will encourage quick and accurate computation. Daily use of mathematical games is an excellent source for improving computational fluency (Kamii & Rummelsburg, 2008). The important factor in practice is that students are engaged and practice occurs frequently so that difficult problems can be identified and deficits addressed.

Multiplication Using Number Lines

The basic process of multiplication is shown using number lines. The student begins at zero and jumps to the right; the first factor tells how many jumps, and the second factor tells the group within each jump. Students who struggle may have difficulty with this model, and the CRA/CSA sequence can assist. It is suggested that students are familiar with the grouping model above in order to assist with remediation of their skills associated with the number line. Without a firm understanding of repeated addition and the translation of numbers and symbols into groups of a particular number, a number of error patterns may arise as students attempt to move up the number line. The concrete level involves placing manipulative items next to each numeral written on a number line. It might be helpful to write the number line on a small individual whiteboard or piece of paper so that marks can be made to show that the objects are grouped. The problem, two times four, would be two groups of four. The student would jump two times up the number line, once from zero to the numeral four, and twice from four to eight. The representational level would include the same process using a tally next to each numeral on a number line. The concrete and representational levels provide the student with an understanding of how the movements up the number line are structured according to the numbers and symbols within a multiplication problem. An example of this process is shown in Figure 6–7.

CRA/CSA APPLICATION ONE-DIGIT MULTIPLIERS AND REGROUPING

When beginning instruction in multiplication of larger numbers, students should have an understanding of the operation, combining or joining groups of equal size. Problems with one-digit multipliers such as 3 × 14 can be

Concrete Model for Problem 2 × 3

Place an object next to each numeral on the number line.

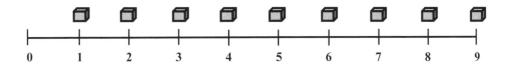

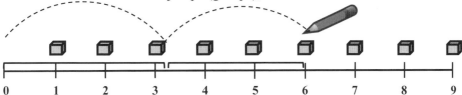

Representational Model for Problem 2 × 3

Place an object next to each numeral on the number line.

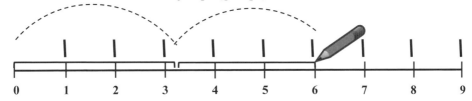

Figure 6–7. *Concrete- and Representational-Level Models Using the Number Line.*

solved using different approaches, but each one requires that students know the meaning of numbers (14 = one ten and four ones) and that multiplication involves combining groups of equal size (3 × 14 = three groups of one ten and three groups of four ones). To help students build understanding, numbers should increase from single-digit numbers to single times a multiple of 10 to single times a larger number that isn't a multiple of 10. Students should be challenged at each step to try to make sense of the problems and solve them. For example, if students have mastered single-digit times single-digit numbers, providing the student with a problem of 3 × 10 and asking how this could be solved (using words, pictures, manipulatives, or equations) can provide insight into where students may or may not need support. Students in need of interventions may understand these concepts if instruction at previous levels was successful, but some students may have gaps in their understanding. The use of CRA/CSA can assist in reminding students of their previous learning or in bolstering their conceptual knowledge, filling in gaps. This section shows how CRA/CSA can be used to teach students the use of partial products, partial products using the area model, and the traditional/standard algorithm.

CRA/CSA and Partial Products

CRA/CSA can be used to support students as they solve problems that involve numbers larger than 10 (23 × 46 = 3 × 6 + 20 × 6 + 40 × 3 + 20 × 40 = 1,058). The use of partial products decreases the likelihood of errors made within the shortened algorithm related to regrouping numbers (e.g., forgetting to add the extra 10 or hundred to the product of the numbers multiplied or forgetting to note that regrouping occurred). The use of partial products involves fewer procedures or shorthand notations. It also connects more naturally to the prior knowledge most students bring to double-digit multiplication. However, solving multiplication with regrouping problems in this way still requires that students understand the properties of numbers and operations. Using the CRA/CSA sequence provides students with physical models and representations of these concepts. Using manipulatives and pictures, the composition of each number within the problem is shown (23 = 20 + 3 and 46 = 40 + 6) as well as the multiplication concept (3 groups of 6, 20 groups of 6, 40 groups of 3, and 20 groups of 40). All of the partial products are added to together to compute the final product. At the concrete and representational/semi-concrete levels, using the partial product approach can be made easier by using a place value table. A simple table can be constructed to assist students in organizing base 10 blocks when solving problems, and an example is shown in Figure 6–8. Examples of partial products problems solved at the concrete and representational levels are shown in Figure 6–9.

CRA/CSA and the Array Model Combined With Partial Products

Solving problems using the partial products approach can be modeled using arrays, showing each partial product as a smaller portion of a larger area. This is often the first model introduced when teaching multiplication of larger

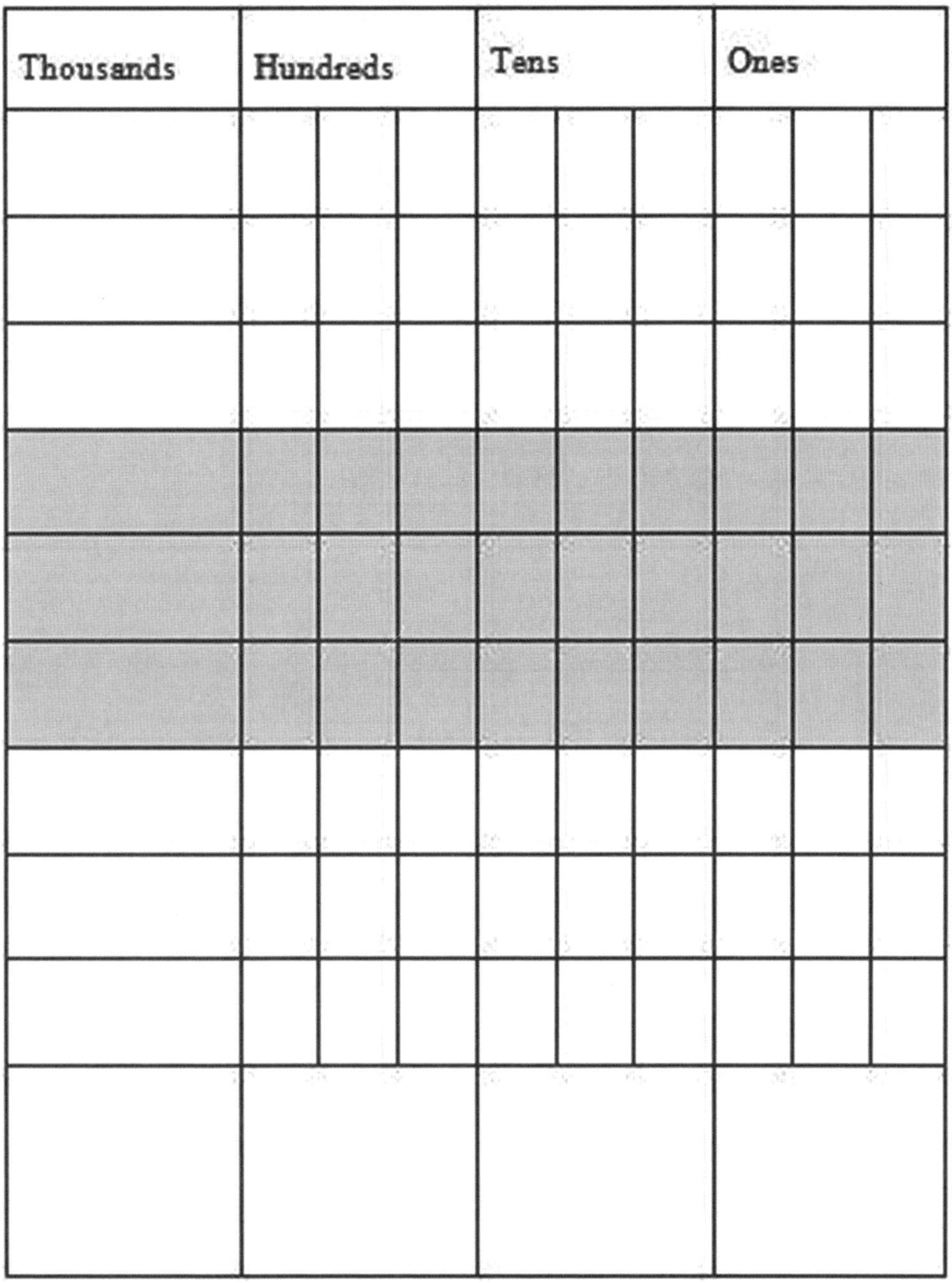

Figure 6–8. Multiplication Mat.

numbers because the visual connects with the visual for single-digit multiplication using the area model. Within the concrete level, for the problem 23 × 5, 23 is broken into its component parts and each is multiplied by 5 (20 × 5 and 3 × 5). Using a grid, multiplication of each number is shown by making

Problem: The students ate 4 boxes of cookies and each box had 36 cookies. How many cookies did the students eat?

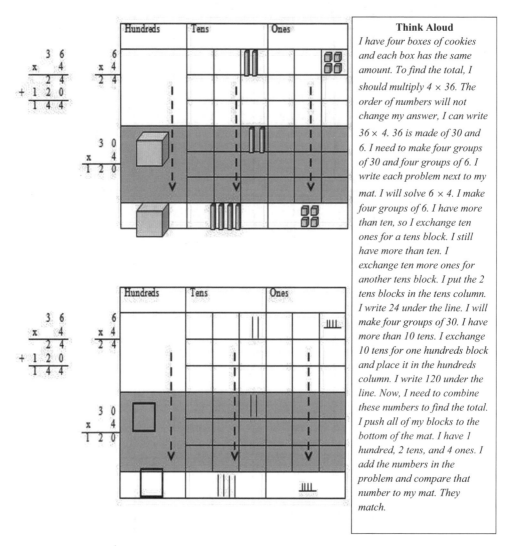

Figure 6–9. Example of Concrete- and Representational-Level Processes for Solving Problems Using the Partial Product Approach.

groups with columns (five columns to represent five groups) and filling the columns with the appropriate amounts. Solving this example at the concrete level would entail filling five columns each with two base 10 tens blocks, and below that, the five columns would be filled with three base 10 ones blocks. Solving this problem at the representational level would involve shading each of the partial products. The processes for concrete and representational instruction are shown in Figure 6–10.

Problem: There are 23 classrooms in the school and each classroom has five computers. How many computer are in the school?

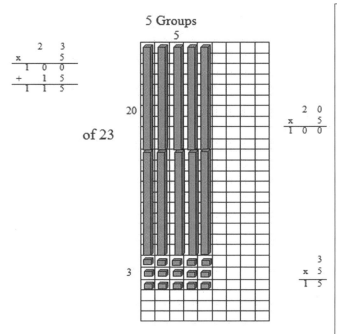

Think Aloud
There are 23 classrooms and each has the same number of computers. To find the total, I should multiply 23 × 5. 23 is made of 20 and 3 and I need to make five groups of 20 and five groups of 3. I write each problem next to my grid. I will solve 20 × 5. I count five columns and put 2 tens blocks in each column. I count them all. I have 10 tens or 100. I write that under the line in the problem. I will solve 3 × 5. I count five columns and put 3 ones blocks in each column. I count them all. I have 15. I write that under the line in the problem. Now, I need to combine these numbers to find the total. I have 1 hundred and 15 ones. I add the numbers in the problem and compare that number to my grid. They match.

Figure 6–10. *Concrete- and Representational-Level Computation Using Partial Products and the Area Model. continues*

CRA/CSA and Traditional Algorithm

As with earlier multiplication concepts, CRA/CSA instruction involves the use of manipulatives to solve problems at the concrete level, drawings within the representational/semi-concrete level, and finally problem solving using numbers only with the aid of a mnemonic strategy. Researchers have found that CRA/CSA instruction is more efficient when students have assistance in organizing manipulatives and drawings; manipulation of larger numbers can become cumbersome if students do not have a structure for organizing numbers (Flores & Franklin, 2014; Flores, Schweck, & Hinton, 2014). A place value table has been a successful tool used by students to solve problems using base 10 blocks and drawings; an example that could be used when one factor is nine or less is shown in Figure 6–11.

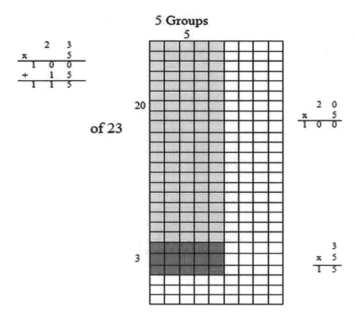

Figure 6–10. continued

	Hundreds			Tens			Ones	

Figure 6–11. Place Value Table.

The simple table shown in Figure 6–11 is used to assist students in organizing base 10 blocks or drawings. Within each column (hundreds, tens, ones), there are nine cells. These cells are used to form groups so that manipulatives or drawings remain organized.

Instruction at the concrete level involves solving a problem using base 10 blocks and a place value table printed on a mat, large enough for the manipulatives. Instruction begins with reading the problem, attending to the mathematical symbol and the numbers in the problem; the example described shows the process for solving 24 × 3. The multiplicand (the number being multiplied) is represented using base 10 blocks, and the teacher should point out the composition of the number (24 is made with two tens and four ones). The base 10 blocks are set on the place value mat in the appropriate columns (two tens in one cell within the tens column and four ones in a cell within the ones column). With attention to the multiplier (the number multiplied by), groups are made beginning in the ones place. There are four ones and, according to the problem, there should be three groups of four. Groups of four ones blocks are placed in each of three cells in the ones column. The total number of blocks is determined; if the product is 10 or more, regrouping occurs. It is important to question students and ensure they can explain why regrouping would be needed and can show it at the concrete stage before moving on to the representational/semi-concrete stage.

A simple rule for this assessment is, "If there are 10 or more, go next door." However, this rule needs to be understood and explained by students. The mention of this simple statement is not meant to bypass the importance of conceptual understanding. Students who struggle with mathematics are a diverse group of students, which includes students with language-processing deficits that manifest in a variety of learning problems. Providing students with a memory aid in the form of a rhyme will assist them in retrieving information associated with their previous learning and understanding. Simply teaching students a rule, without conceptual understanding, will result in further failure.

They need to be able to share that leaving a double-digit number in the ones place changes the value of the number as a whole. Since three groups of four is 12, 10 ones blocks are exchanged for one tens block; the tens block is placed in the tens column. This example illustrates the need for the student to understand that 12 ones is the same as 1 ten and 2 ones. If students struggle with this concept during multiplication, the content about place value needs to be revisited. It is helpful to place this block above the grouping cells, horizontally. This placement provides a visual cue that the additional 10 is not part of the original number. Regrouping is noted within the written problem by writing a small numeral 1 over the tens place. After regrouping, the remaining ones are counted and that number is written within the problem. Next, multiplication occurs for the tens and the multiplier. There are two tens and the multiplier is three; two tens blocks are placed in each of three cells. The total number of tens blocks, including the block above that was regrouped, is combined and that number is noted in the written problem. A step-by-step example of this process is shown in Figure 6–12.

Representational/semi-concrete instruction follows the same process using drawings rather than base 10 blocks. Ones are represented using small vertical tallies drawn on a horizontal line (as shown in basic multiplication models in the previous section). Tens are represented using long vertical lines. Hundreds are represented using squares. An example of the process is shown in Figure 6–13.

Problem: There are three third-grade classrooms and each classroom has 24 students. How many third-grade students in all?

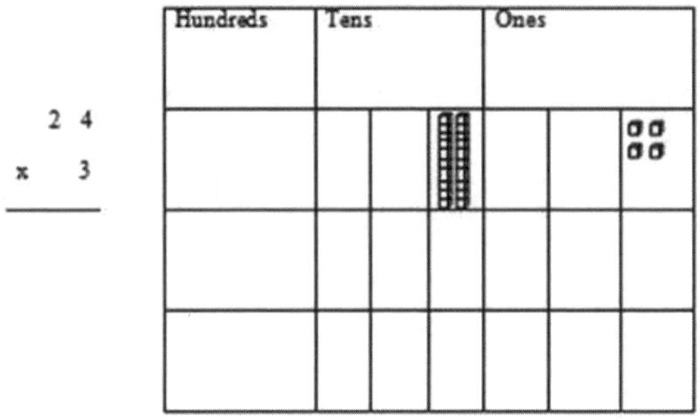

	Hundreds	Tens	Ones
2 4		[1 ten rod]	[4 ones]
× 3			

Think Aloud

There are three classrooms and each has the same amount of students. I am combining groups of students and each group has the same amount: 3 × 24. I know that the order of the numbers does not change the answer, so I can set up the problem as 24 × 3 so that it will be easier to solve. I begin by making 24, 2 tens and 4 ones. I place the tens in one section of the tens column and I place the ones together in a section of the ones column.

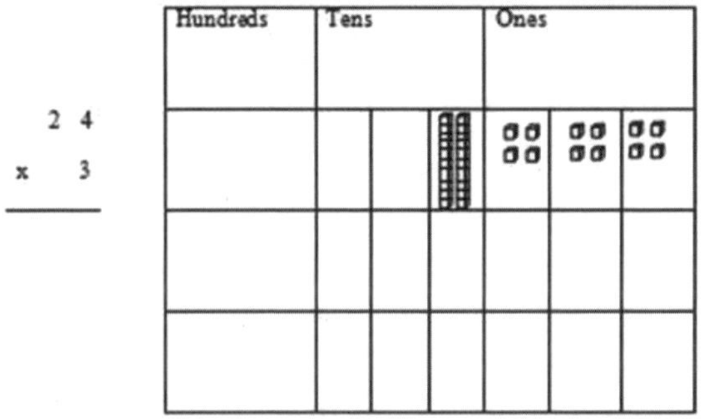

	Hundreds	Tens	Ones
2 4		[1 ten rod]	[4][4][4]
× 3			

Think Aloud

I begin in the ones column. I have four in each group and I need three groups. I have one group and I will make two more groups for a total of three.

Figure 6–12. Concrete-Level Instruction Using Traditional Algorithm to Solve Multiplication With Regrouping (One-Digit Multiplier). continues

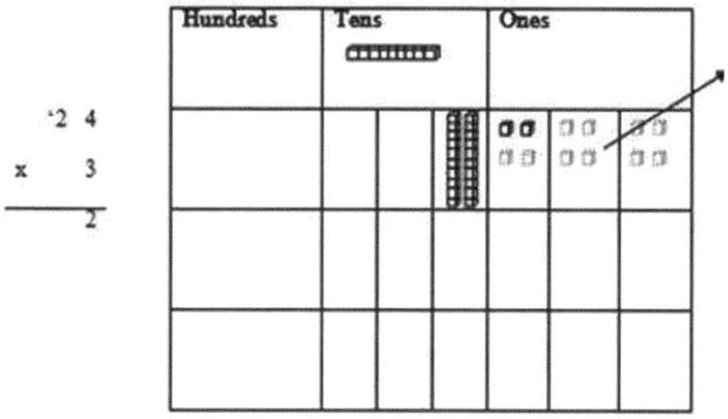

¹2 4
× 3
───
    2

Think Aloud

Three groups of 4 is 12. If I have 10 or more, I go next door. I exchange 10 ones for a 10. I make sure I clear those 10 ones off my mat so that I am not confused later. I put the tens block in the tens column, turning it sideways to remind me it is an extra 10 that I will add later. I mark my problem, noting the 10. I have two ones, so I write 2 under the line.

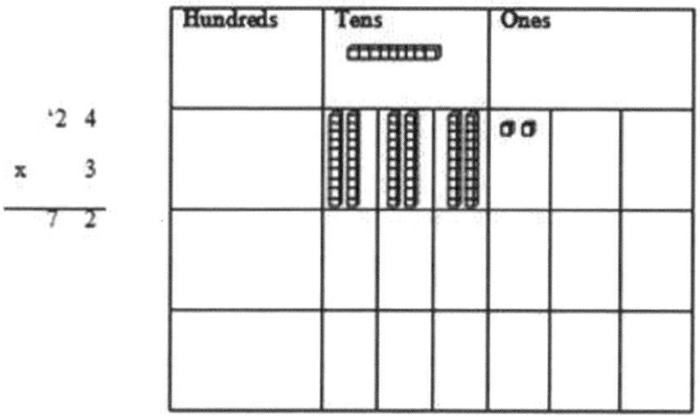

¹2 4
× 3
───
 7 2

Think Aloud

Three groups of 2 tens is 6 tens and I add the extra 10. I have 7 tens altogether. I mark my problem and write 7 under the line in the tens place. I need to check my problem. Does my mat match the numbers in the problem? Yes.

Figure 6–12. continued

Problem: There are three teams and each team has 24 players. How many players in all?

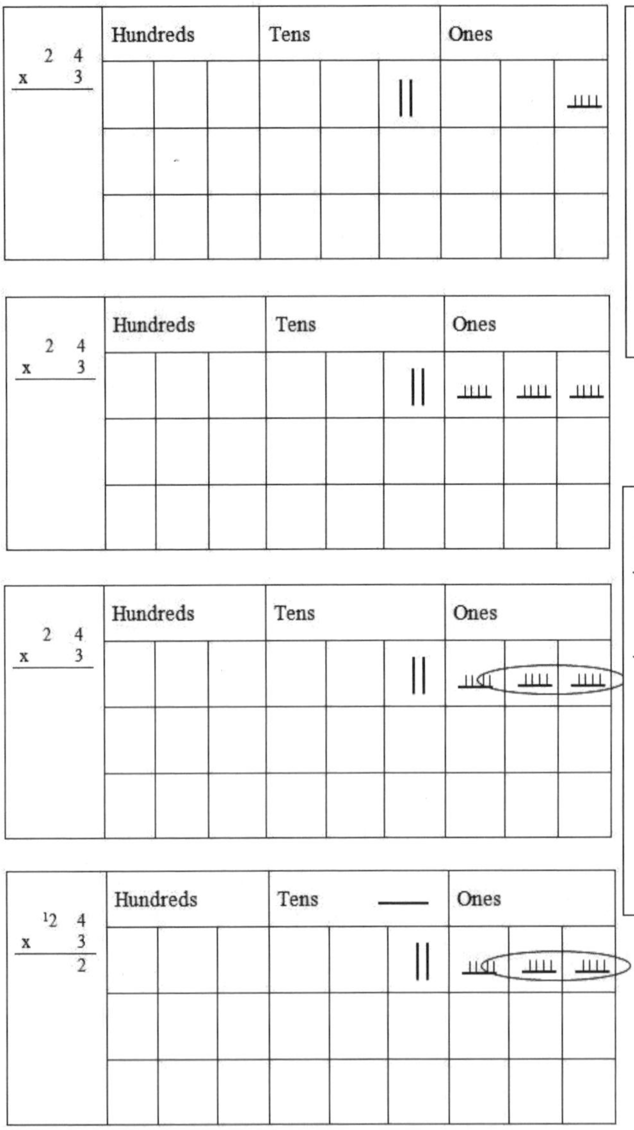

	Hundreds	Tens	Ones
2 4			
x 3		||	‖‖‖‖

Think Aloud
There are three teams and each has the same amount of players. I am combining groups of players and each group has the same amount: 3 × 24. I know that the order of the numbers does not change the answer, so I can set up the problem as 24 × 3 so it will be easier to solve. I begin by drawing 24, two tens drawn in the tens place.

	Hundreds	Tens	Ones
2 4			
x 3		||	‖‖‖‖ ‖‖‖‖ ‖‖‖‖

	Hundreds	Tens	Ones
2 4			
x 3		||	(‖‖‖‖) ‖‖‖‖ ‖‖‖‖

Think Aloud
I begin in the ones column. I have four in each group and I need three groups. I have one group and I will draw two more groups for a total of three.

Three groups of 4 is 12. If I have 10 or more, I go next door. I circle 10 ones and draw a 10 sideways in the tens column to remind me it is an extra 10 that I will add later. I mark my problem, noting the 10. I have two ones, so I write two under the line.

	Hundreds	Tens	—	Ones
¹2 4				
x 3		||		(‖‖‖‖) ‖‖‖‖ ‖‖‖‖
2				

Figure 6–13. *Representational-Level Instruction Using Traditional Algorithm to Solve Multiplication With Regrouping (One-Digit Multiplier).*

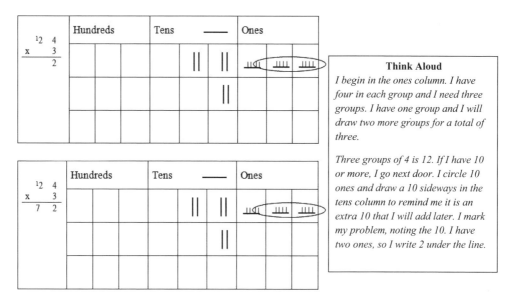

Figure 6–13. continued

Abstract instruction involves solving problems using numbers only. Research has shown that a strategy taught between representational/semi-concrete and abstract instruction has been effective in assisting students with the problem-solving procedures (Flores & Franklin, 2014; Flores, Hinton, & Strozier, 2014; Flores, Schweck, & Hinton, 2014). However, it is critical that students demonstrate conceptual knowledge prior to teaching the strategy. The strategy follows: (a) read the problem, (b) examine the ones, (c) note the ones, (d) address the tens, (e) mark the tens, and (f) examine the hundreds and exit with a quick check (RENAME). The RENAME strategy provides students with the procedural steps for completing the algorithm, providing students with steps that will lead to efficient computation and attention to details. Steps within the strategy prompt students to act in ways that will decrease the likelihood of frequent error patterns. For example, the first step ensures that students attend to the operation and numbers, the next step ensures that students begin in the ones place, and the final step prompts students to examine their work, checking that the answer is reasonable. The RENAME strategy is taught to mastery prior to abstract instruction, meaning that students can recall each step. During abstract instruction, problems are solved using the RENAME strategy, so it is important that students can effortlessly recall the strategy. An example of a think-aloud demonstration with RENAME is shown in Figure 6–14.

Problem	Strategy	Think Aloud
1 2 3 6 x 4 ――― 1 4 4	**R**ead the problem	The problem is 36 × 4. There are four groups of 36.
	Examine the ones	In the ones place, there are six ones and I will make four groups of 6. 4 × 6 is 24. The rule is, "If there are 10 more, go next door." I have more than 10, so the two tens must be put in the tens place. I write 2 over the tens place.
	Note the ones	Since I put the two tens in the tens place, now I note the ones that are left. I write 4 in the ones place.
	Address the tens	In the tens place, I have three tens and I will make four groups. 4 × 3 tens is 12 tens and I have two more. Altogether, I have 14 tens. The rule is, "If there are 10 more, go next door." I have more than 10, so the 10 tens or 100 must be put in the hundreds place. I write 1 over the hundreds place.
	Mark the tens	Since I put the one hundred in the hundreds place, now I mark the tens that are left. I write 4 in the tens place.
	Examine the hundreds; exit with quick check	I am finished multiplying 36 × 4, but I have 100. I write that under the line in the problem. I check my problem. I made sure that I added all of the numbers that were regrouped before I noted answers under the line. My answer seems reasonable because 30 × 4 would be 120. My answer is close.

Figure 6–14. Think-Aloud Example of Abstract Problem Solving Using RENAME Strategy.

CRA/CSA APPLICATION REGROUPING WITH TWO-DIGIT MULTIPLIERS

When multiplying using larger numbers such as when both factors are two-digit numbers, the computation process becomes lengthy regardless of the approach used. The traditional algorithm involves shortening of the process but with more procedural steps. The partial products approach involves combining more numbers of larger quantities. Both leave room for many errors, so it is essential that students have conceptual understanding and the ability to estimate for reasonableness of answers. As mentioned previously, for students to master this type of complex computation, number sense and a firm understanding of place value are needed. This should not be introduced until students are conceptually sound in multiplication with smaller numbers. The CRA/CSA sequence provides students with representations of numbers that serve to remind them of these concepts and further solidify this previous learning. Students without this awareness of numbers and place value within the base 10 system will struggle with complex multiplication because it will become a procedure to be memorized and likely poorly remembered without conceptual understanding as to why procedures are needed and used.

CRA/CSA, Two-Digit Multipliers, and Partial Product Approach

As with the previous description of this approach using one-digit multipliers, it does not involve as many procedures as the traditional, shortened algorithm. Instead, it involves multiplication of larger numbers and addition of three or more numbers to find the final product. For example, the problem 34 × 23 would require the sum of the following products (30 × 3, 4 × 3, 20 × 4, and 30 × 20). The use of a place value mat or table will assist students in organizing objects and drawings. The area model is useful in demonstrating the partial product method so students are able to see the four multiplication problems within a double-digit times a double-digit problem (e.g., 34 × 23).

At the concrete level, instruction would begin by decomposing both numbers within the problem (30 + 4 and 20 + 3). Then number sentences are made with each number (30 × 20, 30 × 3, 4 × 20, 4 × 3). On the place value mat, each of these problems would be solved using base 10 blocks. First, 30 groups of 20 would be made and the total number of tens would be combined into hundreds blocks. The same process for solving this problem with hundreds blocks as described in the previous section can be used. The section underneath, three groups of 30, would be made. No regrouping is required so one would proceed with the next partial product; four groups of 20 would be made in the next section. Finally, four groups of three would be made and the product assessed using the "10 or more" rule; 10 ones are exchanged for a 10 and the tens block placed in the tens column. Finally, all of the base 10 blocks in each column are combined to arrive at the final answer. This is likely to involve additional regrouping. Representational/semi-concrete instruction involves the same procedures, but drawings are used instead of base 10 blocks. An example of partial products is provided in Figure 6–15.

Partial products can also be shown using area models. Rather than using a place value table or mat, base 10 blocks would be placed on a grid as described previously. At the representational/semi-concrete level, the area of a grid would be shaded; it is helpful to use different colors when shading the grid to find different partial products. An example of problems solved at the concrete and representational/semi-concrete levels using an area model is shown in Figure 6–16.

CRA/CSA and the Traditional Standard Algorithm

Solving multiplication problems with two-digit multipliers using the traditional algorithm is best implemented with the use of a place value table, as shown in the previous section. However, the table must be modified in order

Problem: There were 23 bags of candy bars handed out on Halloween. Each bag had 34 candy bars. How many candy bars were handed out on Halloween?

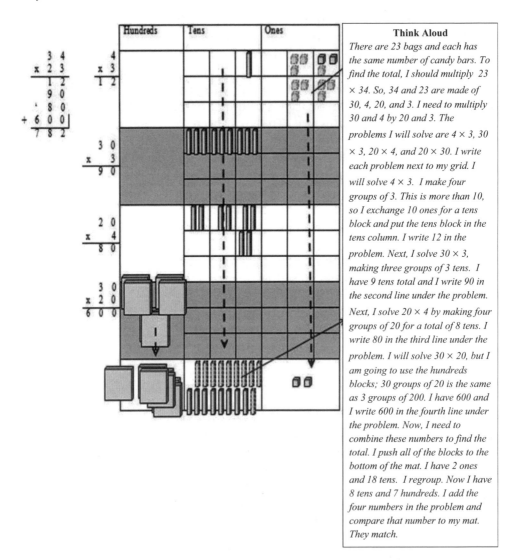

Figure 6–15. Example of Concrete Problem Solving Using Partial Products.

Problem: There are 15 soccer teams in the tournament and each team has 23 players. How many players are in the tournament?

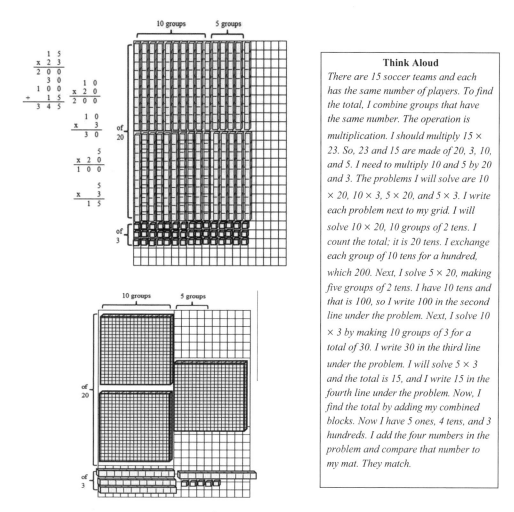

Figure 6–16. Concrete- and Representational-Level Examples of Multiplication With Two-Digit Multipliers Using Partial Products and Area Models. continues

to account for multiplication across both numerals within the multiplier. An example of a multiplication table is shown in Figure 6–17.

The table shown in Figure 6–17 provides students with prompts for solving problems using the traditional algorithm: (a) there are nine cells that can be used for multiplying, consistent with the base 10 system; (b) the shaded area is differentiated since this section is used to manipulate objects or pictures when multiplying by the number in the tens place of the multiplier; and (c) the ones place within the shaded section is missing cells because ones will not be grouped when multiplying by the number in the tens place

Representational Level

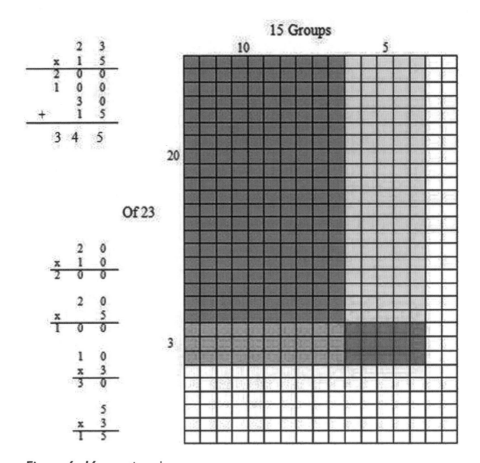

Figure 6–16. continued

Thousands			Hundreds			Tens			Ones		

Figure 6–17. Multiplication Table Used to Solve Problems With Two-Digit Multipliers.

of the multiplier. In addition to an organizational tool, it provides students with prompts that assist in and set the stage for using the procedures that will be the focus of instruction that will occur after the representational phase of instruction.

The steps for solving problems with two-digit multipliers using the traditional algorithm at the concrete level are as follows. The problem used for this description is 34 × 23. The multiplicand (top number) is decomposed and represented on the place value mat using base 10 blocks (34 represented with three tens blocks and four ones blocks, each placed a cell in the appropriate column). Beginning in the ones place, groups of the number within the ones place of the multiplicand are made (three groups of four ones blocks). It is critical that students revisit the need to regroup when there are more than nine in a place value column. Since the product of 3 × 4 is 12, 10 ones blocks are exchanged for one tens block and the tens block is placed in the tens column above the grouping cells. The remaining ones blocks are counted and the number is recorded within the written problem. Moving to the number in the tens place of the multiplicand, groups of the number are made (three groups of three tens blocks). The product is added to the one 10 that was previously moved to the tens place and the sum is evaluated using one's knowledge of the number system; if there are 10 or more, one must regroup. Since the sum of 9 + 1 is 10, 10 tens blocks are exchanged for one hundreds block and the hundreds block is placed in the hundreds column. The number of remaining tens is noted, zero, because there are zero tens. The zero is written in the tens place of the written problem and the hundreds place is examined, noting that there is one hundred; one is written in the hundreds place within the written problem. This rule for regrouping must be understood conceptually, rather than memorized by students. It is essential that they can explain why 10 cannot remain in ones place value column.

Next the process begins again; however, the multiplicand is multiplied by the number in the tens place of the multiplier. Since multiplication using the number in the ones place of the multiplier is finished, this number is crossed out and a zero is written underneath the previous row of answers in the ones place. On the place value mat, objects are manipulated in the tens place. The numbers to be multiplied are 20 and 4. This problem could be represented by making 20 groups four or four groups of 20. It is important to think aloud during the problem-solving process and model efficient problem solving and point out concepts along the way. The answer to the problem of 20 groups of four and four groups of 20 will be the same; this is the commutative property that students learned within beginning multiplication lessons. It would be easier and therefore more efficient to make four groups of 20 because the alternative, 20 groups of four, will result in 80 ones blocks, which will be grouped together to form eight tens blocks. Make this decision and the thinking process obvious to students. So, four groups of two tens blocks are made and the product evaluated; eight tens or 80 are less than 10, so eight is written in the tens place within the written problem.

Next, the number in the tens place of the multiplicand is multiplied by the tens place of the multiplier, 30 × 20. This is another instance in which thinking aloud and number sense instruction are critical. Within the problem, the product of the numbers 20 and 30 is a number within the hundreds place. The teacher should explicitly point out that numbers will be manipulated in the hundreds column on the place value mat. The teacher should think aloud, wondering about how that might be. The teacher makes 20 groups of three tens or 30. Once the groups are made, resulting in 60 tens blocks, group the tens blocks to make six hundreds blocks. The teacher shows that 20 × 30 is 600. The teacher shows the students that there is a way to compute this problem using hundreds blocks rather than tens blocks, two groups of three hundreds blocks or two groups of 300. An appropriate understanding of numbers and place value involves knowing that numbers in the hundreds place are 10 times larger than numbers in the tens place, and conversely, numbers in the tens place are 10 times smaller than numbers in the hundreds place. Therefore, 20 × 30 is the same as 2 × 300. This step in the process provides for another opportunity to remind students or provide remedial instruction in this concept. It is likely to require several exposures to this concept, so continue the process of multiplying using tens blocks and then hundreds blocks across several problems. After arriving at the answer using the base 10 blocks, the product is evaluated using understanding about "10 or more." Since 600 is less than 10 tens or 1,000, the 6 can be recorded in the hundreds place within the written problem. Next the columns can be added to arrive at the final answer. The base 10 blocks in the ones column, tens column, and hundreds column are combined and moved to the bottom of the mat. This process may involve regrouping, and if so, manipulate the blocks accordingly. The numbers within the written problem are added and the final written answer is compared to the blocks represented at the bottom of the place value mat. The process for solving this problem is shown in Figure 6–18.

Instruction at the representational/semi-concrete level follows that same process; however, drawings are used to represent numbers rather than base 10 blocks. Regrouping is represented by circling drawings to show that numbers have been exchanged. It is suggested that teachers avoid erasing and use a simple efficient method such as circling. An example problem solved at the representational level is provided in Figure 6–19.

After mastery of problem solving at the representational level, the RENAME strategy is taught. The strategy's application is similar to that used to solve problems with one-digit multipliers, but the strategy is applied twice for two-digit multipliers. The strategy is as follows: (a) read the problem, (b) examine the ones, (c) note the ones, (d) address the tens, (e) mark the tens, (f) examine the hundreds and begin again the tens column, (g) read the problem, (h) examine the tens, (i) note the tens, (j) address the hundreds, (k) mark the hundreds, and (l) add the numbers and exit with a quick check. Abstract instruction involves application of the RENAME strategy and solving

6. Teaching Multiplication Using the Concrete-Representational/Semi-Concrete–Abstract Sequence   **155**

Problem: There were 23 cases of juice boxes bought for the school festival and each case had 34 juice boxes. How many juice boxes were bought for the festival?

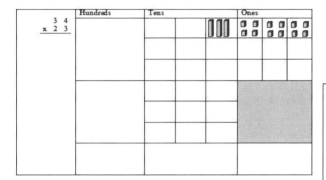

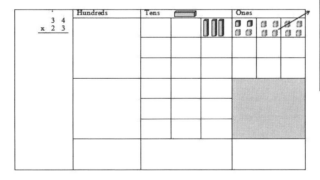

Think Aloud
I begin in the ones place, making three groups of 4. Three groups of 4 is 12. If I have 10 or more, I go next door. I exchange 10 ones for a 10. I make sure I clear those 10 ones off my mat so that I am not confused later. I put the tens block in the tens column, turning it sideways to remind me it is an extra 10 that I will add later. I mark my problem, noting the 10. I have 2 ones, so I write 2 under the line.

Figure 6–18. *Multiplication With Two-Digit Multiplier Concrete-Level Example.* continues

problems using numbers only. It is critical that students master concepts related to the procedure prior to learning and applying the RENAME strategy. Asking students to memorize procedures without understanding how and why they are used will result in their inability to consistently and accurately solve problems as well and leave them ill equipped to learn more complex applications of mathematics.

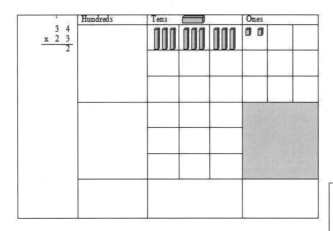

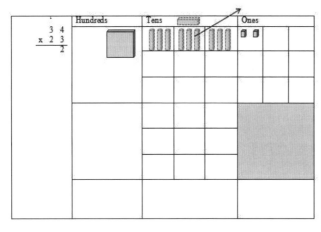

Think Aloud

I have 30 and need to make three groups. I have one group of 3 tens and I need to make two more. Three groups of 3 tens is 9 tens and I add the extra 10. I have 10 groups of 10, so I need to exchange tens for one hundreds block. I put the hundreds block in the hundreds column. I mark my problem, writing zero in the tens place since there are no tens. I write a 1 in the hundreds place. I finish multiplying by the number in the ones place, so cross that 3 out and begin again multiplying by the number in the tens place. Now I am multiplying by 20. I begin a new line and I write zero in the ones place.

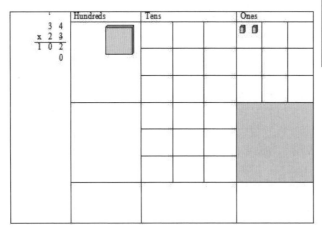

Figure 6–18. continues

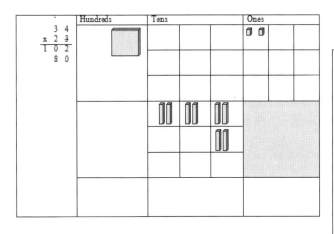

Think Aloud
I have 4 and 20. I am working in the tens place, so I am going to make groups of 20. I make four groups of 2 tens blocks and the total is 8 tens. I mark 8 in the problem in the tens place. Now, I am working in the hundreds place. I am multiplying 30 and 20 and the answer will be a number in the hundreds. Thirty groups of 20 will be a very large amount of tens. I am working in the hundreds place and I know that numbers in the tens place are 10 times smaller than numbers in the hundreds place. I can make this problem easier by using hundreds blocks instead of tens blocks. If I needed 20 groups of 30, I can make the same problem with two groups of 300. Two groups of 300 is 600. I mark the problem with a 6 in the hundreds place.

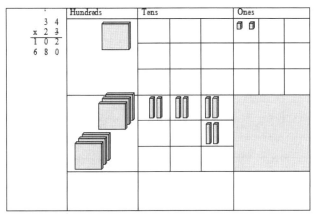

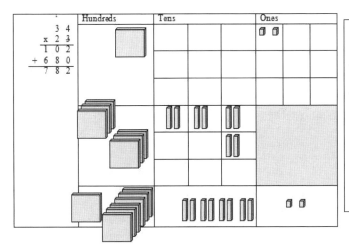

Think Aloud
I am not finished with the problem yet; I need to add my answers to find the total. I push all of the blocks down to the bottom of the mat to find the total. I have 2 ones, 8 tens, and 7 hundreds. I add the numbers in the problem and compare my answers. They match.

Figure 6–18. continued

Problem: There are 34 classrooms in the school and each classroom has 23 students. What is the total number of students at the school?

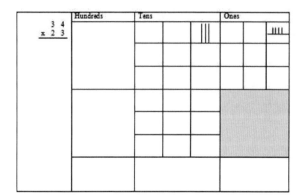

Think Aloud

I have 34 classrooms and each has the same number of students. To find the total, I should multiply 34 × 23. 34 is made of 30 and 4. I draw tens in the tens column and 4 ones in the ones column.

Think Aloud

I draw three groups of 4. I have one drawn so I draw two more. Three groups of 4 are 12.

Figure 6–19. *Multiplication With Two-Digit Multiplier Representational-Level Example.* continues

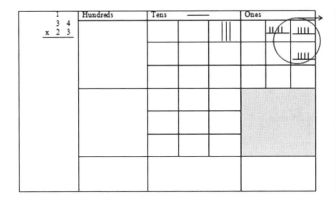

Think Aloud

Three groups of 4 is 12. If I have 10 or more, I go next door. I circle 10 ones and draw a 10 in the tens column sideways to remind me it is an extra 10 that I will add later. I mark my problem, noting the 10. I have 2 ones, so I write 2 under the line.

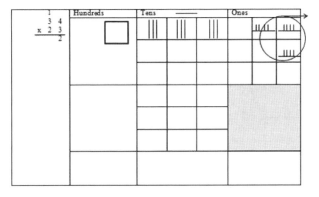

Think Aloud

I have 30 and need to make three groups. I have one group of 3 tens and I need to draw two more. Three groups of 3 tens is 9 tens and I add the extra 10. I have 10 groups of 10, so I need to exchange tens for one hundred. I circle 10 tens and draw one hundred in the hundreds column. I mark my problem, writing zero in the tens place since there are no tens. I write a 1 in the hundreds place.

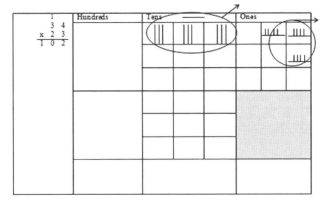

Figure 6–19. continues

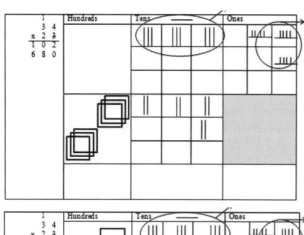

Think Aloud

I am finished multiplying by the ones. I am multiplying by the number in the tens place. I write a zero in the ones place. I have 20 and 4. I am working in the tens place, so I will make four groups of 20. That is 8 tens altogether. I note the 8 in the written problem.

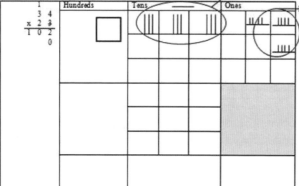

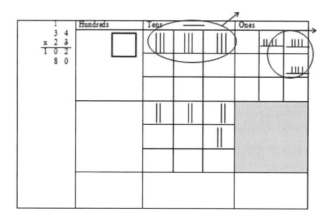

Think Aloud

Next, I multiple 20 and 30. I can complete this problem with hundreds instead of tens so that I can draw fewer pictures. My answer will not change. 20 × 30 is the same as 2 × 300. I make two groups of 300. That is 600 total. I mark the written problem with a 6 in the hundreds place.

Figure 6–19. continues

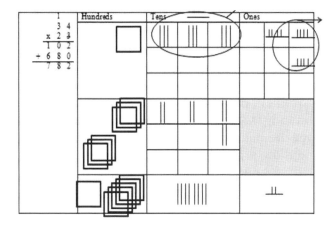

Figure 6–19. continued

PROGRESS MONITORING

Assessing student progress in multiplication is vital to ensuring they are prepared to move forward in the learning trajectory. Progress monitoring and formative assessment should help teachers identify the elements students truly understand, areas of confusion, and areas where they are mimicking procedures taught without conceptual understanding. This knowledge can inform future instructional steps and guide in the progression of this complex topic.

Single-Digit Factor Multiplication

Before introducing single-digit multiplication, teachers need to formatively assess student understanding of place value, value of numbers, joining and combining numbers using addition, ability to create and identify equal groups, and connection between symbols to represent the number of objects. These are all prerequisite skills to multiplication, and additional work on these areas may be needed before students are ready to begin multiplication. If these students are successful with prerequisite skills, teachers can introduce concepts of single-digit multiplication using the CRA/CSA model described above. As students are demonstrating single-digit multiplication at all levels of CRA/CSA, it is important to ask students to solve contextual problems using the specific stage of CRA/CSA, to provide contextual problems using models of various CRA/CSA stages, to have students explain each step and thought involved in solving the multiplication problems, and to ask questions about the multiplier, the multiplicand, and the product. Asking questions that help students explore patterns, such as providing three problems that have a factor of zero and asking students what they notice, is important in guiding students to making generalizations. Asking students about the reasonableness of answers and to predict ranges of possible answers is another step in building conceptual understanding. Also having students relate the area, number line, and set models to equations while sharing relationships and different uses for each type of problem can deepen understanding and help teachers understand student thinking. When using the area model, teachers need to question students about the meaning of rows and columns as well as asking students how they found the product. When using the number line, students need to explain how they knew how many jumps to make, how they decided the length of each jump, and how the product was determined.

One-Digit Multipliers and Regrouping

When progress monitoring for understanding of one-digit multipliers with regrouping, many of the same areas as single-factor multiplication need to be assessed. In addition, it is essential to ensure that students are able to identify how many tens and ones are in the double-digit multiplicand. So asking questions about how 3×14 could be rewritten and expressed is important. Also, having students explain each step in the modeling and why regrouping is or is not needed is essential. It is important to provide problems where regrouping is required and where it isn't, so they don't assume it always should or should not occur. Asking questions about the place value in problems (such as 3×10 in 3×14) to ensure they understand the 1 is one group of 10 will be important as they build upon this stage.

Double-Digit Multiplication and Beyond

When introducing double-digit multiplication, the area model is a way for students to see the four multiplication problems within a two-digit times two-digit problem or the six steps within a two-digit times a three-digit problem. As students use the area model, it is important to ask them what each digit means and to explain what they are doing and how this relates to single-digit multiplication. When students regroup, students should be able to explain why they are regrouping and what the number represents. Students need to be able to explain what a zero means as a place value when regrouping is used and why regrouping is necessary. It is also important that students are able to explain why all problems are added together at the end to find the product (e.g., $34 \times 23 = (30 \times 20) + (30 \times 3) + (4 \times 20) + (4 \times 3)$). It is important to ask contextual questions and to ask questions in a variety of ways to ensure students are not repeating what they have been told without conceptual understanding.

CHAPTER SUMMARY

The purpose of CRA/CSA instruction is to provide students with conceptual understanding of mathematical operations, which is accomplished through solving problems using objects at the concrete level and drawings at the representational/semi-concrete level. Students who struggle in mathematics need more practice or more explicit instruction to master concepts related to numbers and operations, and CRA/CSA provides these. This chapter demonstrated how multiple approaches to multiplication instruction can be made more explicit using the CRA/CSA sequence. It describes the various steps of complexity that students encounter as they build multiplicative concepts.

As multiplication becomes more complex, there is an increase in the procedural knowledge needed to solve problems. Research has shown that the combination of CRA/CSA with the strategic instruction model is effective in leading students to efficient problem solving. The strategies combined with CRA/CSA within this chapter, DRAW and RENAME, provide students with a set of efficient, generalizable steps to assist students in their procedural skills, while also helping to maintain and build conceptual understanding. It is important that instruction in these strategies only be provided when students have demonstrated conceptual knowledge of the operation.

Mathematics standards adopted by most states across the nation require that students explain their problem solving; there is a focus on how answers are found, rather than just accuracy. CRA/CSA and its use of objects and drawings allow students to physically manipulate and see the workings of operations as well as describe what they see. Teachers can help students acquire the language to describe mathematical processes and operations by thinking aloud during demonstrations and asking that students think aloud during guided practice and independent practice. The CRA/CSA sequence provides multiple exposures at each level of instruction, fading visual aids from concrete to representational or abstract instruction; this provides students with scaffolding needed to become more proficient in their description of the problem-solving process. This chapter showed how multiple approaches to multiplication can be made explicit using CRA/CSA. It is important that students understand numbers and operations as well as know that problems may be solved in different ways. Deep conceptual understanding of numbers and operations allows students to think mathematically and choose alternate approaches. The CRA/CSA sequence provides students who struggle with mathematics with the tools necessary to become mathematical thinkers so that computation is a problem-solving and thinking process rather than a set of rules and procedures that students use without thought or understanding.

Finally, it is critical that students who struggle with mathematics have experience using multiplication within authentic contexts. Instruction should include word problems that require multiplication as well as other operations that have been mastered previously. Understanding operations involves both knowing how to use the operation as well as when to use the operation. In summary, multiplication instruction should emphasize conceptual understanding, application within authentic situations, and procedural knowledge. Using the CRA/CSA sequence across instructional approaches can assist students in mastering rigorous mathematical standards, becoming mathematical thinkers and efficient problem solvers.

APPLICATION QUESTIONS

1. How are number sense and understanding of place value related to conceptual understanding of multiplication?

2. Using any approach, how would instruction in beginning multiplication be taught at each of the levels of the CRA/CSA sequence?

3. What is the rationale for using place value tables or mats when teaching multiplication with regrouping; how do these tools benefit students?

4. Provide reasons why one might choose each of the following approaches to multiplication with regrouping: traditional algorithm, partial products, or area models.

5. How would a teacher think aloud and provide a rationale for choosing to solve a word problem using multiplication rather than addition or subtraction?

REFERENCES

Common Core State Standards Initiative (CCSSI). (2010). *Common Core State Standards for Mathematics*. Washington, DC: National Governors Association Center for Best Practices and the Council of Chief State School Officers. Retrieved from http://www.corestandards.org/assets/CCSSI_Math%20Standards.pdf

Dacey, L., & Drew, P. (2012). Common core state standards for mathematics: The big picture. *Teaching Children Mathematics, 18*, 378–383.

Flores, M. M., & Franklin, T. M. (2014). Teaching multiplication with regrouping using the concrete-representational-abstract sequence and the strategic instruction model. *Journal of American Special Education Professionals, 6*, 133–148.

Flores, M. M., Hinton, V. M., & Strozier, S. D. (2014). Teaching subtraction and multiplication with regrouping using the concrete-representational-abstract sequence and strategic instruction model. *Learning Disabilities Research and Practice, 29*, 75–88.

Flores, M. M., Schweck, K. B., & Hinton, V. M. (2014). Teaching multiplication with regrouping to students with learning disabilities. *Learning Disabilities Research & Practice, 29*(4), 171–183.

Kamii, C., & Rummelsburg, J. (2008). Arithmetic for first graders lacking number concepts. *Teaching Children Mathematics, 14*(7), 389–394.

Mercer, C. D., & Miller, S. P. (1992a). Teaching students with learning problems in math to acquire, understand, and apply basic math facts. *Remedial and Special Education, 13*(3), 19–35.

Mercer, C. D., & Miller, S. P. (1992b). *Strategic math series: Multiplication facts 0–8.* Lawrence, KS: Edge Enterprises.

Miller, S. P., & Mercer, C. D. (1993). Using data to learn about concrete-semiconcrete-abstract instruction for students with math disabilities. *Learning Disabilities Research & Practice, 8*(2), 89–96.

CHAPTER 7

Teaching Division Using the Concrete-Representational/ Semi-Concrete–Abstract Sequence

OVERVIEW

This chapter shows how the concrete-representational/semi-concrete–abstract (CRA/CSA) sequence is used to teach division, from basic facts through division of large numbers. Throughout the chapter, instruction in the division operation will be described along the continuum of students' development of conceptual understanding, procedural knowledge, and declarative knowledge. At each stage of learning, teaching methods using concrete objects, drawings, and procedural strategies will be demonstrated and modeled for the reader. Since learning standards require that students think about mathematics in a flexible manner, completion of the division operation will be described and shown using different approaches. Within each approach, the reader will learn how to teach students using the CRA/CSA sequence with objects (concrete level), drawings (representational/semi-concrete level), and using numbers and procedural strategies (abstract level). The purpose of this chapter is to provide the reader with a rationale for using the CRA/CSA sequence to teach division, as well as methods for implementing the CRA/CSA sequence across various computation approaches.

SEQUENCE OF DIVISION INSTRUCTION WITHIN MATHEMATICS STANDARDS

After students have a firm understanding that objects can be made into equal groups and combined to form larger numbers, the foundation for division can be built. Conceptual understanding of division begins with partitioning a collection of objects into smaller groups of the same size. For example, when given a group of 12 objects, one asks how many groups of three can be made from the whole. Standards begin with the manipulation of objects, partitioning whole groups into small groups of equal size. The mathematical language and symbols associated with division are taught using objects, pictures. Since division instruction follows mastery of basic multiplication, students learn to solve basic division problems using their knowledge of multiplication. A division problem can be solved as a multiplication problem that is missing a multiplier (10 ÷ 2 is the same as 2 × ? = 10). Next, standards that include larger numbers and one-digit divisors follow but do not require a particular method or algorithm. Rather, students use their knowledge of operations, properties, and numbers to solve problems using pictures and objects. The same approach is taken when learning about division with two-digit divisors. Use of the standard algorithm for long division is not included in many elementary-level standards and is addressed at the middle level. As the standards are written, students master concepts and use their conceptual understanding to learn procedural knowledge. The standards (CCSSI, 2010) related to division are located in Table 7–1.

DESCRIPTION OF DIVISION AND PREREQUISITE SKILLS

The division operation builds on students' previous learning of subtraction, the separation of numbers. However, in division, numbers are separated in equal groups. This notion of breaking an amount into equal groups goes hand in hand with students' understanding of multiplication, the joining of groups of equal size. A prerequisite to division is an understanding that numbers can comprise equal-sized groups and the same number can be added repeatedly. Just like the relation between addition and subtraction, multiplication and division are inverse operations. Teaching students multiplication conceptually and following that with conceptual instruction in division will highlight this inverse relationship. After conceptual understanding is established, and the objective of instruction is fluency, students' knowledge of multiplication assists in their learning of division facts through the notion of fact families (3 × 2 = 6, 6 ÷ 3 = 2).

Once students master basic facts, division of larger numbers requires prerequisite skills related to understanding of the base 10 system, the decomposition and composition of numbers, and multiplication. In completing division problems involving large numbers, students must have a firm understanding of multiplication related to regrouping, including its prerequisite skills. Students must understand that the number 427 is four hundreds, two tens, and seven ones and that this number can also be made using 42 tens and seven ones, as well as other combinations of numbers. Without this understanding, breaking the number 427 into 36 groups (427 ÷ 36) will be quite difficult. The use of the CRA/CSA sequence can be applied throughout the developmental process of teaching division to ensure that these prerequisite skills are bolstered and firm prior to the learning stage at which

Table 7–1. Division Standards

> Interpret whole-number quotients of whole numbers; for example, interpret 56 ÷ 8 as the number of objects in each group when 56 objects are partitioned equally into eight groups, or as a number of groups when 56 objects are partitioned into equal shares of eight objects each.
>
> Use multiplication and division within 100 to solve word problems in situations involving equal groups, arrays, and measurement quantities (e.g., by using drawings and equations with a symbol for the unknown number to represent the problem).
>
> Determine the unknown whole number in a multiplication or division equation relating three whole numbers.
>
> Apply properties of operations as strategies to divide.
>
> Understand division as an unknown-factor problem. *For example, find 32 ÷ 8 by finding the number that makes 32 when multiplied by 8.*
>
> Fluently divide within 100, using strategies such as the relationship between multiplication and division (e.g., knowing that 8 × 5 = 40, one knows 40 ÷ 5 = 8) or properties of operations.
>
> Find whole-number quotients and remainders with up to four-digit dividends and one-digit divisors, using strategies based on place value, the properties of operations, and/or the relationship between multiplication and division. Illustrate and explain the calculation by using equations, rectangular arrays, and/or area models.
>
> Find whole-number quotients of whole numbers with up to four-digit dividends and two-digit divisors, using strategies based on place value, the properties of operations, and/or the relationship between multiplication and division. Illustrate and explain the calculation by using equations, rectangular arrays, and/or area models.

procedural strategies are taught. Students who struggle in mathematics often fail to achieve conceptual understanding, which then causes great struggle when procedural strategies are taught. Students who struggle cannot execute procedures without knowledge of how or why they are necessary; therefore, they need intensive experiences to form conceptual understanding. The use of CRA/CSA and explicit instruction can provide these experiences.

CRA/CSA APPLICATION FOR BASIC DIVISION

Mercer and Miller (1992) developed and field tested materials in which the CRA/CSA sequence was used to explicitly teach basic operations, including division. The CRA/CSA sequence was used to model division using objects, drawings, and numbers. At the concrete level, division was taught using objects; a group of objects was separated into equal groups. At the representational level, the same process of separating was implemented with drawings. At the abstract level, students learned a procedural strategy and used this to complete simple division equations using just numbers. The procedural strategy used for division is the same as the one used for other basic operations, and its purpose is to assist the student in being attentive and thoughtful while completing any basic operation. The strategy asks the student to (a) discover the sign, (b) read the problem, (c) answer or draw and check, and (d) write the answer (DRAW).

Teaching Division by Separating Equal Groups

Instruction in division is a natural extension of basic multiplication instruction. With multiplication, students develop an understanding that a number can be repeatedly added to make a larger number (2 + 2 + 2 + 2 = 8 is four groups of two). That same understanding is extended with the inverse as a number is repeatedly separated from a larger number until one reaches zero or until one cannot continue to subtract (8 − 2 − 2 − 2 − 2 = 0 is eight is separated into four groups of two). The mathematical symbols are translated into words; the expression 6 ÷ 2 means six can be separated into how many groups of two? Concrete-level instruction teaches this translation through the physical manipulation of objects. For the example problem of 6 ÷ 2, a group of six objects is presented along with plates or other grouping items and objects; two objects are successively removed from the larger group and put on a plate. This problem is complete when all six of the objects have been removed and distributed across three plates. This process demonstrates that six comprises two groups of three or 6 ÷ 2 = 3. Given this problem example, the quotient (3) tells how many objects groups are made by the divisor (2) after the dividend (6) is systematically separated in groups equal to the divisor. When teaching division, present computation problems within the context of real-life situations. From the inception of the student's conceptual understanding of the division operation, ensure that its application is clear. Students who struggle with mathematics may also struggle with language; therefore, provide opportunities for translation and practice throughout each stage of learning. The process for concrete division instruction is shown in Figure 7–1.

After students use manipulatives to solve basic division problems with consistent accuracy (80% of answers to problems completed independently are correct), instruction moves to the representational level. This involves the use of drawings rather than objects to represent the manipulation of numbers within the operation. Drawings should be simple; avoid elaborate shapes or caricatures that distract the student from the task of completing the

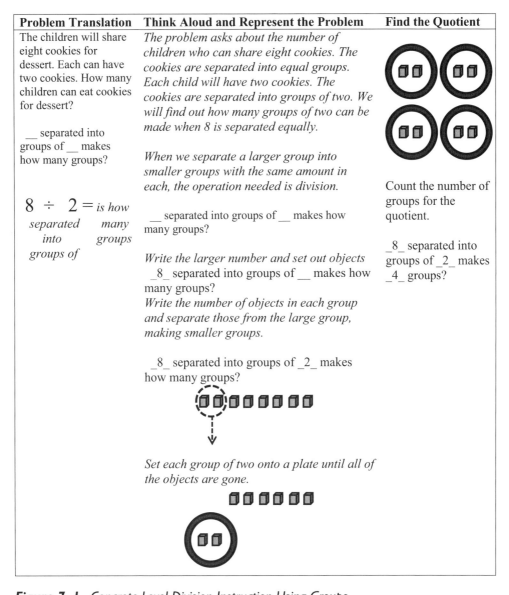

Figure 7–1. Concrete-Level Division Instruction Using Groups.

operation. Simple drawings are exemplified by short tallies and horizontal lines to represent groups.

Drawings continue to provide the student with a visual aid that shows how numbers are changing due to the operation. Since the student draws and physically shows the separation of a number into groups, he or she can better attribute meaning to the operation as well as express his or her understanding in words. At the representational level, the student now has two experiences in which the operation is modeled and conceptual understanding of the operation is firm when the student can reliably and consistently describe the operation while solving a problem. The presentation of problems within context at the representational level provides repeated opportunities to teach and use language, translate words into mathematical symbols, and assign meaning to the operation. Continued practice in solving applied problems assists the student in forming a schema for division and the situations in which it is used. Instruction at the representational level is shown in Figure 7–2.

Problem Translation	Think Aloud and Represent the Problem	Find the Quotient
The teacher has a box of 12 markers. Each group of students should be given three markers for a project. How many groups can have enough markers for their project? __ separated into groups of __ makes how many groups? $12 \div 3 =$ *is how separated many into groups groups of*	*The problem asks about the number of student groups that will each have three markers. The 12 markers are separated into equal groups of 3. We will find out how many groups of 3 can be made when 12 is separated equally.* *When we separate a larger group into smaller groups with the same amount in each, the operation needed is division.* __ separated into groups of __ makes how many groups? *Write the larger number and draw the larger number using tallies.* _12_ separated into groups of __ makes how many groups? *Write the number of objects in each group and separate those from the large group, making smaller groups.* _12_ separated into groups of _3_ makes how many groups?  |||||||||||| *Draw a horznital line under each group of three until all of the tallies are gone.*  |||| ||||||||	 ||||| ||||| Count the number of groups for the quotient. _12_ separated into groups of _3_ makes _4_ groups.

Figure 7–2. Example of Representational-Level Division Instruction.

After students have demonstrated mastery of division at the representational level, instruction moves to the abstract level in which no visual aids are provided. Mercer and Miller developed a procedural strategy discussed in previous chapters to assist students in completing problems at the abstract level, the DRAW strategy. To review, the DRAW strategy represents the following procedural steps: (a) discover the sign, (b) read the problem, (c) answer or draw and check your answer, and (d) write the answer. The execution of these steps prompts the student to be attentive during the problem-solving process by attending to the operational symbol as well as attending to the numbers when reading the problem. The third step provides the student with another option for discovering the answer by drawing if the student has not committed the answer to memory. Finally, the fourth step prompts the student to write the answer; this addresses potential problems students face when they draw the problem and become distracted from the original problem and forget to write the numerical symbol once they have drawn the representation of the completed problem.

Within the abstract phase, the focus of instruction is development of automaticity. Although students rely on the DRAW strategy, it is important that their approach to answering involves less drawing and increasingly more automatic recall of answers. Within the abstract phase, encourage students to use their knowledge of multiplication facts to recall division facts. For example, if given the problem $8 \div 2$, recall the multiplication fact that involves those numbers ($2 \times ? = 8$). Instruction at the abstract level should involve practice activities that encourage speed and accuracy. Include fluency practice in the form of games to supplement drill and practice activities in order to encourage and maintain students' interest and motivation.

DIVISION WITH REMAINDERS

Once students understand the concept of division as the process of separating numbers into equal-sized groups, introduce the notion of remainders. Frequently, numbers cannot be separated across groups evenly; there are some that remain and another whole group cannot be made. For example, $11 \div 2$ comprises five groups of two with one left ungrouped. The presence of a remainder is shown in one of two ways. The first way is to leave the number as it is, showing that there are more left ($11 \div 2 = 5$ and remainder of 1). Another way is to express the remainder as a fraction, splitting the remainder or showing it as a proportion, having one out of the needed two to form a group ($11 \div 2 = 5\frac{1}{2}$). Instruction at the concrete level explicitly shows this situation of having more than can be grouped evenly as well as how and why the notation of a remainder is expressed. An example of instruction involving division with remainders at the concrete level follows in Figure 7–3.

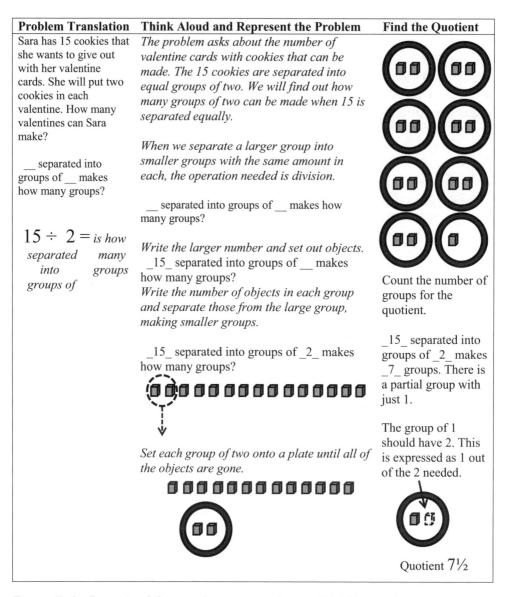

Figure 7–3. Example of Concrete Instruction in Division With Remainders.

Instruction at the representational involves the use of drawings. Simple drawings consist of vertical tallies and horizontal lines used to show grouping. An example of division with remainders taught at the representational level is shown in Figure 7–4.

Students must also understand the impact of remainders in real-life situations. Present students with contexts in which the presence of a remainder impacts the problem-solving process and the actions needed. In real-life

Problem Translation	Think Aloud and Represent the Problem	Find and Make Decisions About the Quotient
Four children shared a bag of candy. There were 25 candies in the bag and the four children shared the same amount. How much candy did each child receive? __ separated into groups of __ makes how many groups? $25 \div 4 =$ *is how separated many into groups groups of*	*The problem asks about the number of candies that each child can have if they share candies. To find the answer, we separate 25 using equal groups of 4. The number of groups we make will tell us how many candies each child can have.* *When we separate a larger group into smaller groups with the same amount in each, the operation needed is division.* __ separated into groups of __ makes how many groups? *Write the larger number and draw the tallies.* _25_ separated into groups of __ makes how many groups? *Write the number of objects in each group and separate those from the large group, making smaller groups.* _25_ separated into groups of _4_ makes how many groups?  ||||||||||||||||||||||||| *Underline each group of 4 until all of the tallies are used or there are less than 4.*  |||| |||| |||| |||| |||| |||| |	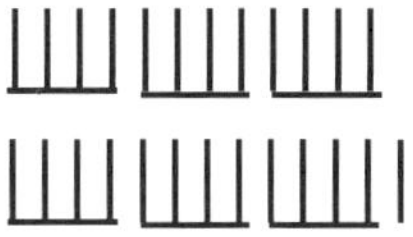 |||| |||| |||| |||| |||| |||| | Count the number of groups for the quotient. _25_ separated into groups of _4_ makes _6_ groups. There is a partial group with just 1. The partial group of 1 should have 4.  | The group of 1 should have 4. This is expressed as 1 out of the 4 needed. Quotient 6¼

Figure 7–4. Example of Representational Instruction in Division With Remainders.

situations, a remainder may force us to round to the next whole number; if a class needs four and one-fourth boxes of pencils, the teacher will purchase five boxes in order to provide pencils for all. In other situations, the remainder may be disregarded. If a group of students were to share 35 cookies, telling how many students could receive four cookies would involve leaving the remainder out of the answer; eight students could each have four cookies. Conceptual understanding of remainders includes knowledge about how remainders affect problem-solving situations such as the previous examples as well as decision making related to how remainders are used. Instructional lessons should include thinking aloud in which the teacher models decision making regarding the impact of remainders on the answer to the problem-solving situation. Examples of this type of instruction follow in Figure 7–5.

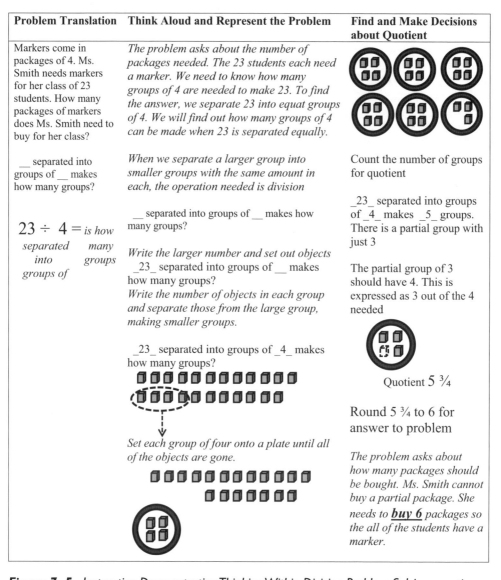

Figure 7–5. Instruction Demonstrating Thinking Within Division Problem Solving. continues

DIVISION OF LARGE NUMBERS

As students make progress in their understanding of the operation of division, they will encounter larger numbers. Understanding as well as proficiency in multiplication is a critical prerequisite to completion of different approaches to division using large numbers. All approaches to division of large numbers are dependent upon efficient and reliable multiplication skills. Instruction should proceed in a manner that ensures that students understand how numbers are manipulated within the chosen division algorithm; therefore, instruction at the concrete level should be taught first. The same way in

Problem Translation	Think Aloud and Represent the Problem	Find and Make Decisions About the Quotient
Keisha is making bags of cookies for a bake sale. There should be three cookies in each bag. Her mom made 35 cookies for the bake sale. Keisha's mom said that she can eat the cookies left over that do not fit in bags. How many cookies can Keisha eat? __ separated into groups of __ makes how many groups? $35 \div 3 =$ _is how separated many into groups groups of_	*The problem asks about the number of cookies that will be left after making packages. The 35 cookies will be separated into groups of 3. To find the answer, we separate 35 into equal groups of 3 and will see if there is a partial group.* *When we separate a larger group into smaller groups with the same amount in each, the operation needed is division.* __ separated into groups of __ makes how many groups? *Write the larger number and set out objects.* _35_ separated into groups of __ makes how many groups? *Write the number of objects in each group and separate those from the large group, making smaller groups.* _35_ separated into groups of _3_ makes how many groups?  *Set each group of 3 onto a plate until all of the objects are gone.* 	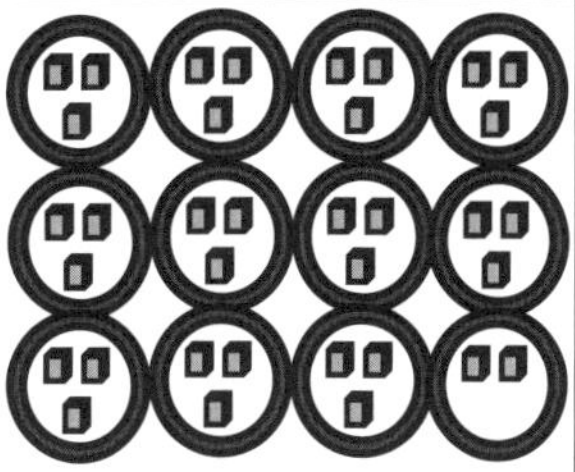 Count the number of groups for the quotient. _35_ separated into groups of _3_ makes _11_ groups. There is a partial group with just 2. The partial group of 2 should have 3.  Quotient 5 r2 The remainder, 2, is the answer to the problem. *The problem asks about cookies left over that cannnot be make into packages. Even though Keisha can make five equal packages, the problem asks how many are left over. She has **2 left** that cannot make a complete pack.*

Figure 7–5. continued

which smaller numbers are systematically separated into groups according to the divisor, the same concept should be taught using larger numbers. This process is shortened using different algorithms; however, concrete instruction should occur first. An example of concrete instruction using large numbers is shown in Figure 7–6.

One way to approach division of large numbers is through the partial quotients algorithm. This algorithm involves the use of prior knowledge of multiplication in which the student separates portions of the dividend according to the divisor, keeping track of partial quotients until the dividend is depleted. The partial quotients are added together to arrive at the

> There are 120 students in the fourth grade. There are 15 students in each fourth-grade classroom. How many fourth-grade classrooms are there?
>
> *The problem asks about how many classrooms or groups of students. The fourth-grade students are grouped into classrooms with 15 students in each. The total number, 120 students, is separated into groups of 15. When a number is separated into groups of the same size, the operation is division.*
>
> $120 \div 15 =$ *is how many groups?*
> *separated into groups of*
>
> $15 \overline{)120}$
>
> The number, 120, is represented below. 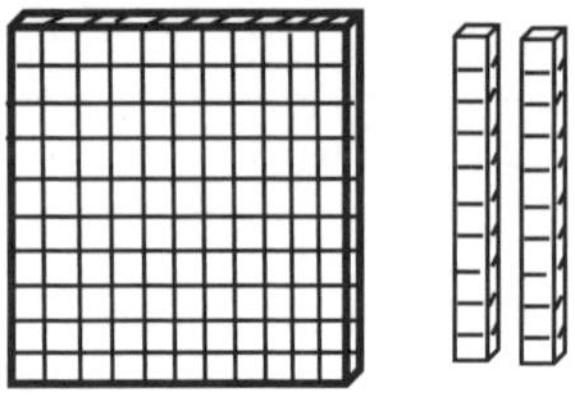
>
> In order to make groups of 15, the total number must be separated below. 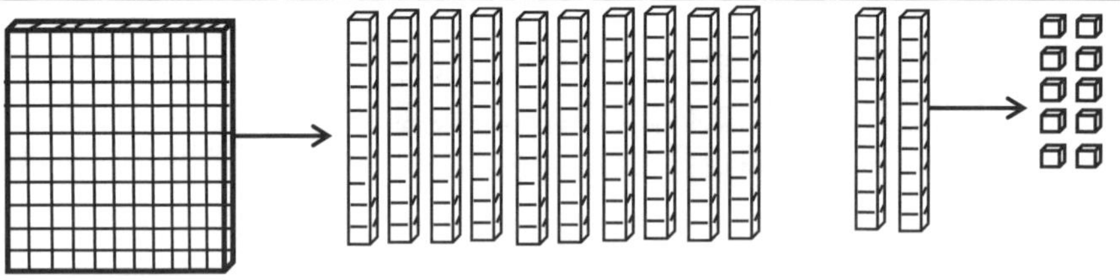
>
> The number of groups of 15 is counted to find the quotient of 8.

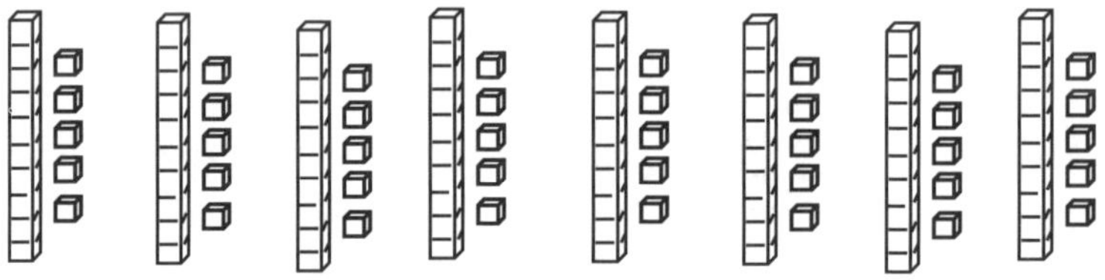

Figure 7-6. Concrete Instruction of Division With Large Numbers.

final quotient. This approach provides flexibility in making the quotient; the student can choose a variety of combinations of multiplication problems involving the divisor. An example of instruction in partial quotient is shown at the concrete level in Figure 7–7.

The traditional algorithm can be more difficult for students since the manipulation of numbers is hidden behind the procedural steps. Furthermore, the shortened steps are unlike and more difficult than other shortened algorithms in the following ways: (a) the symbol used to show division differs

There are 195 students in the fourth grade. There are 15 students in each fourth-grade classroom. How many fourth-grade classrooms are there?

The problem asks how many classrooms or groups of students. The fourth-grade students are grouped into classrooms with 15 students in each. The total number, 195 students, is separated into groups of 15. When a number is separated into groups of the same size, the operation is division.

195 ÷ 15 = *is how*   15 ⟌195
  separated   many groups?
  into groups of

Using partial quotients, find how many groups of fifteen can be made. The process of separation begins in the hundreds place. The hundred is broken into tens. Now there 19 tens. There are 15 tens (10 x 15) within 195. So, I note ten to the side. I subtract 150 from 195. Now, observe how many groups of 15 can be made with the 45 that are left. Three groups of 15 can be made. Add the partial quotients to find the answer. Add 10 and 3 for an answer of 13.

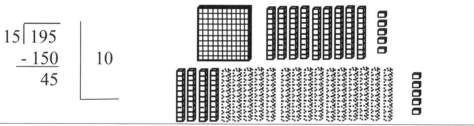

Now, observe how many groups of 15 can be made with the 45 that are left. Three groups of 15 can be made (break a ten into ten ones to make groups of 15). Add the partial quotients to find the answer. Add 10 and 3 for an answer of 13.

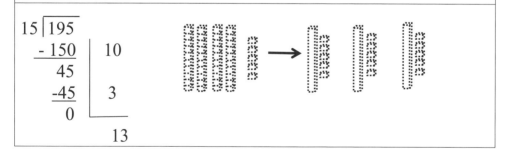

Figure 7–7. Concrete-Level Instruction in Division Using Partial Quotients.

since the ÷ symbol is replaced with the half-rectangle encased around the dividend, (b) problem solving begins in the left-hand column rather than in the ones place on the right, and (c) the separation of the dividend can become lost with an emphasis on multiplication in finding the quotient. Teaching students to use this algorithm should begin with conceptual instruction that shows how and why the shortened procedures are used. Students should be able to observe the process of separation of the dividend. The authors do not advise that teachers use the full CRA/CSA sequence to teach division of large numbers since that would be cumbersome, but instead, show students how and why the shortened processes work using concrete demonstration. The following shows instruction using the traditional algorithm at the concrete level within Figure 7–8.

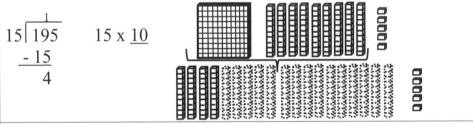

Figure 7–8. Concrete-Level Instruction in the Traditional Algorithm.

CHAPTER SUMMARY

The operation of division might be characterized as the most difficult of the four basic operations in that it requires conceptual understanding of a combination of other operations: separation as well as grouping since division is separation of equal groups from a larger number. In addition to this understanding, the procedural knowledge required to execute various algorithms involves multiple steps. Students who struggle in mathematics may have weak number sense and understanding of subtraction and multiplication operations; this combination of deficits makes learning division, especially division of large numbers, particularly difficult. Students in this situation are left to execute procedures without understanding, relying on rote memory, which will likely result in failure and frustration. Therefore, it is critical that students learn about division at the conceptual level using concrete and representational instruction prior to being asked to solve problems at the abstract level. This chapter showed concrete instruction using division algorithms to solve problems using large numbers. The authors emphasize the importance of this type of instruction because the conceptual jump from $6 \div 2$ to $187 \div 16$ is too great; students can easily become distracted and caught up in procedures. An emphasis on the conceptual underpinnings of each algorithm must be internalized by the student prior to computation using numbers at the abstract level.

APPLICATION QUESTIONS

1. How are number sense, understanding of place value, and operations related to conceptual understanding of division?

2. Using any approach, how would instruction in beginning division be taught at each of the levels of the CRA/CSA sequence?

3. Provide reasons why one might choose to teach division using partial quotients.

4. Develop a word problem that a teacher could use to show how division is relevant to students' experiences.

REFERENCES

Common Core State Standards Initiative (CCSSI). (2010). *Common Core State Standards for Mathematics*. Washington, DC: National Governors Association Center for Best Practices and the Council of Chief State School Officers. Retrieved from http://www.corestandards.org/assets/CCSSI_Math%20Standards.pdf

Mercer, C. D., & Miller, S. P. (1992). Teaching students with learning problems in math to acquire, understand, and apply basic math facts. *Remedial and Special Education*, *13*(3), 19–35.

CHAPTER 8

Understanding Fractions Using the Concrete-Representational/Semi-Concrete–Abstract Sequence

OVERVIEW

This chapter illustrates how students struggling with concepts related to fractional understanding can be supported by helping them connect the multiple representations of fractions using the concrete-representational/semi-concrete–abstract (CRA/CSA) sequence. While it is important that students are able to fluidly move between these representations, beginning with the concrete level can help students more fully understand the abstract representation. In addition, encouraging students to make connections between various concrete and semi-concrete/representational models, such as the area, set, and length models, also strengthens the conceptual understanding that will be important as computation of fractions is introduced.

This chapter addresses beginning fraction concepts, comparing fractions, ordering fractions, and fraction equivalency. Operations with decimals and fractions will be addressed in the next chapter. These are very complex and abstract concepts for most elementary students (Sowder & Wearne, 2006). Even many adults, including individuals majoring in elementary education, struggle with concepts related to fractions (Bezuk & Bieck, 1993; Tirosh, Fischbein, Gracher, & Wilson, 1998). This chapter explains how to build conceptual understanding of fractions beyond simply recognizing the shaded portion of a shape. Readers will learn how the CRA/CSA sequence supports students along the learning trajectory for fraction concepts. CRA/CSA processes will be shown using various models of fractions and instructional approaches related to fractional concepts. This chapter also provides readers with a rationale for choosing the CRA/CSA sequence to provide instructional support for students who struggle with fractional concepts as well as guidance for implementing CRA/CSA.

SEQUENCE OF FRACTION INSTRUCTION

Fraction instruction begins in early grades. Students partition shapes and explore equal shares. However, fractions usually become a major focus when students enter Grade 3. In third grade, students represent fractions using area models and line diagrams using denominators beyond just two (half) or four (fourths), as might have been explored in earlier grades. With modeling of fractions using area and line diagrams, students explore the concept of equivalence and comparison. At this stage, students use physical models, pictures, and drawings to identify fractions that are equal or to compare fractions to other numbers. Students also learn about composition and decomposition related to fractions, meaning that three fourths comprises one fourth plus one fourth and another fourth. Development of these fraction concepts is critical in understanding fractions as numbers and forms the foundation for operations and other complex tasks.

DESCRIPTION OF FRACTIONS AND PREREQUISITE SKILLS

Experiences with fractions should begin when students are young. Fractions take understanding of numbers a step beyond the understanding of whole numbers. It involves understanding the elements between whole numbers. Fractions are always defined relative to the whole. The whole may be an area, amount, or a length. Student mistakes concerning fractions often involve inappropriate reasoning of whole numbers. Students often identify the two digits in a fraction as two whole numbers, rather than a single number representing a proportional value (Saxe, Gearhart, & Seltzer, 1999). For example, students may see 3/4 and recognize the whole numbers three and four without understanding that these particular digits are used to represent a number that is less than one. This example does not imply that fractions are numbers that are all less than one; students may see 4/3 and recognize the whole numbers three and four without understanding that these particular digits are used to represent a number that is more than one. While these are common mistakes, this inappropriate whole-number reasoning is not inevitable. With proper instruction and experience, student understanding of the number system will extend beyond whole numbers.

Fractions have multiple meanings, which is one reason they can be so difficult for some children. Therefore, it is important to explore fractions as part of a whole region, fractions as part of a set, and fractions as measures or length. To understand fractions in any of these models, students must be able to identify the whole, understand equipartitioning, and interpret the visual models.

Identifying the Whole

In order to work flexibly and make connections between fractions, it is essential that students understand how the two digits in the fractional number interact with each other to make one value. However, many students struggle with interpreting a fraction as one value (Petit, Laird, Marsden, & Ebby, 2016). In fact, this novice conception of interpreting a fraction as two whole numbers can often be inadvertently encouraged in the classroom when teachers ask for the number that is the numerator and the number that is the denominator or when they discuss least common multiples in exploring equivalent fractions. These types of conversations encourage students to focus on the individual digits in the fraction, rather than seeing the fraction as a whole or exploring the relationship between the digits to create one value. When students do not understand fractions as a single value but instead focus on the two digits as two distinct numbers, they have difficulty comparing, operating, and finding parts of fractions (Petit et al., 2016). Students need experiences with concrete and visual models to understand fractional concepts. They also need experiences ordering, comparing, and estimating fractions. Unitizing is a prerequisite skill to understanding fractions. Being able to recognize subgroups within groups is called unitizing. This skill helps students see the parts contained within a whole in a fraction.

Equipartitioning

It is essential that students have firm conceptual understanding of whole-number concepts and operations with whole numbers in order to grasp the concepts of fractions of these whole numbers. Equipartitioning is the dividing of an area into equal regions. Equipartitioning helps students with concepts such as fair shares, unit fractions, comparing fractions, equivalency, and operations with fractions (Confrey et al., 2010). In order to effectively equipartition, students need to create equal-sized parts, groups, or areas. They also must be able to manipulate within the whole in order to create the correct number of parts. Finally, students must use the entire whole or collection of items (Confrey, Maloney, Wilson, & Nguyen, 2010). Understanding the equipartitioning in fractions is a critical prerequisite skill for later understanding, and students should have experiences that build a firm conceptual grasp by partitioning using various models.

Unit Fractions

Another essential concept in understanding fractions is the notion of unit fractions. A unit fraction is one piece of a fractional component. For example, 1/2, 1/3, 1/4, and 1/57 are all unit fractions because they represent one unit of the fractional piece of the whole. Students need opportunities to explore unit fractions because this concept assists students in forming the notion of comparison and ordering fractions with like numerators. Furthermore, students will find it easier to compare fractions with unlike denominators if they have experiences comparing and exploring various representations of unit fractions. For example, when comparing fractions such as 4/5 and 4/6, students reason that 4/5 is one unit fraction away from one whole and 4/6 is two unit fractions away from one whole. Therefore, exploration of unit fractions involves the concept that 4/5 represents four unit fractions of fifths (1/5 + 1/5 + 1/5 + 1/5). Unitizing can be explicitly addressed using the CRA/CSA sequence through making fractions with manipulatives or drawings and writing the corresponding fractional addition equation.

VISUAL MODELS

Fractions are a very abstract concept for many students. Models are an integral part of understanding fractions and provide a tool to understand the concept of parts of a whole. As students move along the learning trajectory, they need meaningful experiences with models to foster understanding with each concept related to fractions (Petit et al., 2016). Introducing students to various models, including visual models, helps offer different perspectives and deepens full understanding of these concepts (Siebert & Gasking, 2006). These models serve as a means to support students as they move toward abstract thought and understanding. In order for students to be flexible in their use of fractions, they need to experience various ways to represent fractions and transition between the representations in a variety of situations (Lesh, Landau, & Hamilton, 1983). We will describe how these models can be shown using the CRA/CSA sequence to help develop fractional concepts with students.

Area

This is the most common model used when teaching fractions. It involves a whole area that is equipartitioned and the fraction is identified through a means such as shading or filling with concrete objects. Often circles, squares, rectangles, and hexagons are used when exploring the area model of fractions, because they are easily equipartitioned and are shapes students are very familiar with from geometry. However, any shape that can be equipartitioned can be an area model. An isosceles triangle can be equipartitioned into halves or thirds. A five-sided star can be equipartitioned into fifths or tenths. Any shape can be equipartitioned into fractions, but for purpose of teaching students who struggle, it is advised to begin with models that are fairly easily equipartitioned.

Set

The set model demonstrates that fractions are not always parts of one whole object. Instead, fractions can be parts of a whole group of objects. For example, 3/6 could be three of the six girls or 3 of the 6 crayons. This is important, because students often think of fractions as the shaded part of a shape, rather than part of a whole. This can be useful as fractions are expanded to discuss ratios and proportional reasoning in later grades.

Length

The length model can involve fraction bars to explore fractions in terms of length of space; the number line is a length model. This chapter discusses both the bar utilized without marking a number line and then with the number line model, which builds upon students' prior experiences with number lines and whole numbers. The number line model should be used in classrooms because it helps students develop fractional reasoning and makes connections with whole-number conceptual understanding.

Number lines are an important model for students because they help students understand magnitude. Magnitude was an important concept as students developed conceptual understanding of whole numbers. They build on this understanding, using the number line, when fractions are introduced. Fractions differ from whole numbers in that, for any two fractions, there is always another fraction between them. For example, 2/4 and 3/4 are considered to be next to each other on the number line, but 5/8 actually goes between them. The number of fractions that can be placed between two fractions is infinite. This is something that is misunderstood by many children and adults. The number line helps students grasp this concept of density (Saxe et al., 2007). Models including number lines help students develop both sequential and proportional reasoning. Measurement is an important application of number lines that often involve fractions.

CRA/CSA APPLICATION FOR BASIC FRACTIONAL UNDERSTANDING—CONCRETE

CRA/CSA is used as an intervention when students demonstrate they are struggling with fractions. Butler, Miller, Crehan, Babbitt, and Pierce (2003) found that CRA/CSA was the most effective strategy to enhance student understanding of fraction equivalency. Beginning with concrete manipulatives to support students who struggle has been shown to lead to lasting understanding of mathematical concepts (Witzel & Allsopp, 2007). CRA/CSA instruction begins by exploring the concept of fractional parts of a whole (Cramer & Whitney, 2010).

Area

The area model, also sometimes referred to as the region model, is a strong model with which to begin exploring fractional concepts using CRA/CSA. It emphasizes the part-whole relationship and equipartitioning. Rectangles and circles are the most common visuals used for the area model. Both have strengths and weaknesses. The weaknesses appear during the representation phase. Therefore, it is important that students simultaneously explore both visuals. The circle clearly emphasizes the parts of the whole and the whole. However, it is difficult for students to partition a circle, when they move to the representational/semi-concrete phase of CRA/CSA instruction. The rectangle models are easier to partition, but the whole may be difficult to determine when comparing, ordering, and exploring equivalency during the representational phase of CRA/CSA instruction. This is true of the rectangle area model as well as the bar length model. In addition, using concrete models that students are familiar with from other content, such using the hexagon from pattern blocks as the whole, helps students understand that the whole does not just come in the form of a circle or a rectangle.

Intervention instruction begins with discussion about the concept of a fraction. This discussion should include (a) it can be parts of a whole, (b) the pieces need to be equal in area, and (c) one must know how many equal pieces cover a whole in order to determine the fraction displayed when using the area model. Students are then challenged to use pattern blocks to cover shapes to demonstrate parts of the whole. For example, ask students to find how many triangles cover a hexagon and how many trapezoids cover a hexagon. Within this activity, it is important that students are guided to notice how many triangles versus trapezoids cover a shape and their size by specifically prompting, such as, "What do you notice about the size of each piece?" As the teacher and students explore, the teacher will use the language and write the equations for students to see how to represent the concrete fraction symbolically. For example, three rhombi pieces cover one hexagon piece, which represents the whole, so 3/3 = 1 (Figure 8–1).

Another example provides more explicit emphasis on the relation between part and whole in the makeup of the fraction. Present students with

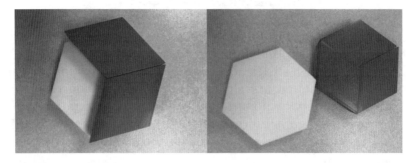

Figure 8–1. *Initial Concrete-Level Instruction of the Area Model of Fractions.*

a shape that has equal parts marked and place blocks within each marked area to make a fraction. Model the thinking processing involved in determining the fraction. Determine how many parts are in the whole and how many of those parts are used. After the teacher models, complete the same tasks together with students in a back-and-forth process in which students identify one portion of the task and the teacher completes the other. Finally, ask students to complete tasks independently and explain their approach to determining the fraction. An example of this is shown in Figure 8–2.

Teacher Model

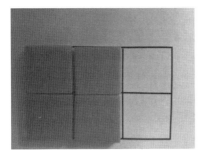

Step 1: Show students the denominator. Teacher thinks aloud while identifying the whole and its component parts.

I have a fraction of a rectangle. The rectangle divided into equal parts. Some are used because they are filled with a block and some are empty. I am going to find the number of parts for the whole rectangle, so I count the ones that are used AND the ones that not used. Count with me, altogether, we have...1..2..3..4..5..6.

This is the denominator, the bottom number of a fraction. I write it below the line. Where should I write 6?

$$\overline{} \\ 6$$

Step 2: Show students the numerator. Teacher thinks aloud while identifying the parts of the whole that are used.

Some of the parts are used because they are filled with a block. I am going to find the number of parts that are used, so I count just the blocks. Count with me, altogether, we have...1..2..3..4.

This is the numerator, the top number of a fraction. I write it above the line. Where should I write 4?

$$\frac{4}{6}$$

Figure 8–2. *Explicit Instruction Identifying Fractions.* continues

Guided Practice

Step 1: Identify the denominator together. Teacher and students think aloud while identifying the whole and its component parts.

We have a fraction of a hexagon. The hexagon divided into equal parts. Some are used because they are filled with a block and some are empty. To find the number of parts for the whole hexagon, we should count the ones that are used AND what else? Yes we need to count the parts that are filled with a block and those that are NOT used. Count all of the parts.

This is the denominator, where do we write this number (prompt might involve pointing). *We write 3 below the line.*

$$\frac{}{3}$$

Step 2: Identify the numerator together. Teacher and students think aloud while identifying the parts of the whole that are used.

Some of the parts are used because they are filled with a block. We are going to find the number of parts that are used, so what should we count? Count with me, altogether, we have...1..2

This is the numerator. Where do we write this number (prompt might involve pointing)? *We write it above the line.*

$$\frac{2}{3}$$

Independent Practice

Present students with another fraction composed of a shape filled with blocks and ask them to think aloud and tell you how they will write the fraction.

Figure 8–2. continued

It is important that the concrete examples begin with fractions that are not more than one, but fractions with value greater than one need to be intertwined as instruction continues in order to avoid the misconceptions that all fractions are less than one. For example, have students find 3/4, 2/3, and then 4/3. This can be challenging but pushes them to examine what the entire fraction means. In order for the area model of fractions to be relevant to students, it is important to also provide real-world scenarios and context where area fractions might occur. For example, a portion of a garden, pizza, or classroom could be used as examples. Figure 8–3 provides an example intervention instructional segment of concrete level with the area model of fractions within real-world contexts.

After students use the fraction manipulatives to solve the fraction problems, instruction moves to the representational/semi-concrete level. However, this chapter continues the discussion of the concrete phase for each of the three models (area, set, and number line) before moving to the representational/semi-concrete and finally the abstract phases. Students need to be exposed to the area, set, and number line concrete models simultaneously before moving to the representational/semi-concrete phase with any of these models.

Set

The set model more clearly connects fractions to multiplication and division. It emphasizes the part-whole relationship and equipartitioning. The challenging concept of explaining how fractions can represent part of a

Problem	Think Aloud and Represent the Problem
A small pizza has eight slices. There are two slices left. What fraction of the pizza is left?	*What does this problem ask? (The problem asks about slices of pizza.) What do we know about a pizza? What can we do to find the fraction of the pizza that is left?*
	Can you use these circle fractions to represent the pizza with all the pieces and then show me the fraction of pieces that are left? Write the fraction of pizza left.
A classroom is divided into four equal areas. One area is the library, one is where student desks are located, one is where the computers are, and the final station is for small group work. What fraction of the room do the computers and library take?	*What does this problem ask? (The problem asks about fractions of a classroom.) What do we know about a classroom? What can we do to find the fraction of the classroom that has computers and the library?*
	Can you use the fraction pieces to represent all classroom spaces? What could you do to show the spaces that are computers and the library? Write the fraction of the room that has computers and the library.

Figure 8–3. Concrete-Level Instruction of the Area Model of Fractions in Context.

group of objects (or people) is something often neglected and overlooked by many educators. Explicitly discussing how this use of fractions is similar and different from the area model is critical for students who struggle with fractional understanding. When working with intervention groups, the set model can be explored at each level of CRA/CSA after the area model has been introduced.

Intervention instruction begins with a review of the meaning of fractions. Begin by reviewing the concept of fractions with students. As students share their conceptual understanding of fractions, use prompts to ensure the following are discussed: (a) a fraction can be parts of a whole, (b) the pieces of a fraction need to be equal in area, and (c) one must know how many equal pieces cover a whole in order to determine the fraction displayed when using the area model. Then, ask students if fractions are always one shape or if a fraction can be found from a group of objects. Pull 12 counters and ask students whether these can be divided in 1/2. After counters are divided into halves, write the fraction 1/2 on a whiteboard and ask students how they knew to make two groups. Ask how many are in each group. Then, explain the basic concepts of the set model, using the set of 12 and an area model to connect the two. Use the concrete area model (such as a hexagon divided in 1/2 with two trapezoids) to connect to the concrete set model. Identify the whole as represented by both models (a hexagon for the area model and 12 counters for the set model). Ask if the division of groups is equal and how the student determined that 1/2 is one of the two groups. Practice with concrete manipulatives should continue until the instructor is confident that the student understands the set model and the expanded view of fractions. It is important that different objects be used during this time. Students, crayons, counters, and numerous other objects that can easily be divided into equal groups may be used (Figure 8–4).

4/5 of the candies are Tootsie Rolls

3/6 of the counters are square

Figure 8–4. *Initial Concrete-Level Instruction of the Set Model of Fractions.*

Continue with the set model and increase the difficulty by asking about sets with larger amounts. For example, present a set of 12 and provide explicit instruction regarding how 3/4 of that set would be determined. An example is provided in Figure 8–5.

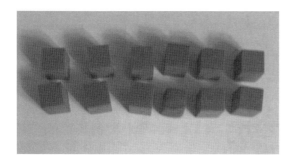

I have 12 blocks. I want to show ¾ of them. First, I notice the denominator or bottom number. This tells me how many equal groups I have. What is the denominator? Yes, it is 4.

So, I divide the whole 12 blocks into four equal groups. I have four cups and put equal amounts in each. Now, I have by 12 blocks divided into four cups.

I am not finished. I have not looked at the numerator, or top number. That tells me how many are used. What is the numerator? Yes, 3. I am going to use three of the four groups to make three fourths.

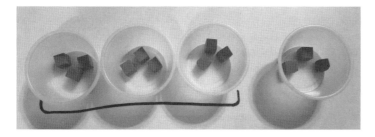

I count all of the blocks in the three groups. Count with me: 3 . . . 6 . . . 9. 3/4 of 12 is how many blocks? It is 9.

Figure 8–5. *Explicit Instruction—Modeling 3/4 of a Set of 12.*

In order for this to be relevant to students, it is important to also provide real-world scenarios and context where area fractions might occur. For example, a portion of people, crayons, paperclips, pencils, books, or candy could be used as examples. See Figure 8–6 for an example of concrete-level instruction within context for the set model of fractions.

Length

As stated previously, the length model may be explored using fraction bars in ways that are similar to the area model with rectangles and other shapes, but the number line is another length model that can help address concepts of

Problem	Think Aloud and Represent the Problem
You have nine candies and want to share the candies equally with two of your friends. What fraction of the candies do you need to give away? How many candies is that? What fraction do you keep? How many candies is that?	*What does this problem ask? (The problem asks you to share nine candies with yourself and two other people.) What do we know about the candies? (There are nine pieces.) How can we use these counters to find out how much each person needs? (Create three piles and give one counter to each pile until you have used all nine.) How many piles do you have? (3) How many are in each pile? (3) If you keep your pile, how many piles did you give away out of the total number of piles? (2/3) How many pieces did you give away? (six of the nine pieces)*
What would it look like if you have a small cake with nine pieces and wanted to share equal amounts of the cake with your two friends? What fraction of the cake do you need to give away? How much cake is that? What fraction do you keep? How much cake is that?	*How is this problem different? The cake is one thing, but the candies were all separate. What is the whole of candies? (nine pieces) What is the whole of the cake? (one cake) Are the numbers within the problem the same? (yes) Do you think the answer will be the same fraction?* *Can you use these circle fractions to represent the cake with all the pieces? What would you do to give equal amounts to the three of you? How much did you give away? Show me the fraction of pieces for both friends and the pieces left for you. How many pieces of the cake did you give away? Write the fraction of the cake you gave away.*

Figure 8–6. Concrete-Level Instruction of the Set Model of Fractions in Context.

magnitude and connect to students' prior knowledge of whole numbers. The bar model enables students to compare the whole to various fractions. Each bar is often a different color and is equipartitioned into a specific amount to allow students to compare how the same whole can be equipartitioned to illustrate various fractions. Being able to place the whole bar on a desk or work mat and then place something such as three of the pieces of the bar that is divided into fourths allows students to see how these are parts of the whole. Students should be encouraged to put bars together to create wholes and then count how many pieces make the whole. For example, students should put the four pieces that make up 4/4 and see that it illustrates the pieces make a length equivalent to the whole fraction bar. Asking students what would happen if another fourth were attached allows them to explore fractions greater than one. Also having students explore equipartitioning bars and creating their own fractions can be useful when the teacher asks students questions about equipartitioning and supports discourse about patterns and structures observed during the partitioning process.

The number line model, also sometimes referred to as the measurement model, supports students' understanding of the density and value of fractions relative to whole numbers. Students use number lines to understand values when learning whole numbers, so learning how fractions fit within this model helps connect fraction meaning to their understanding of whole numbers.

Intervention instruction with the number line model begins by reviewing what a fraction is and the types of fractions models previously explored (area, set, and the bar length model). As students share, guide the discussion using prompts as needed so that the discussion includes the following: (a) a fraction can be parts of a whole, (b) the pieces need to be equal in area or number of pieces, and (c) one must know how many equal pieces make a whole in order to determine the fraction. Students are then shown a number line with zero, one, and two. It is important to make the number line extend beyond one so that students do not only think of fractions represented less than one. It is important to have fraction blocks or bars that match the number line so that students have assistance in equipartitioning. Ask the student to show you where one whole is on the number line and then where two wholes are located. Ensure students understand that the space from zero to one covers one whole space and the space from zero to two covers two whole spaces. Make sure they see the space between one and two cover another whole space. Often students only think of the end point, rather than the value of the entire space as being two. This is as important as a student develops understanding that five apples involves all five apples, not just the fifth one counted.

Now, ask students to identify the point that is halfway between zero and one. Half is a number students are all familiar with from daily use, so it is a nice starting point with all fraction models. Place the half fraction bars or blocks on the number line to see that there are two halves in a whole and to see where 1/2 is on the number line. Using the fraction bar as a guide, students mark the number line. Repeat this with thirds, fourths, sixths, and eighths. Have one page with multiple number lines, so students can see the relationship between the different fractions placed on different number lines such as thirds and sixths, halves and fourths, and eighths, and halves and sixths. This is shown in Figure 8–7.

Once students have worked with unit fractions (e.g., 1/3, 1/5, 1/6), ask them what is represented when you use two of the sixths pieces on the number line; start at zero and mark the space at the end of these (you have 2/6 of one whole). For example, have students show 3/4 with fraction bars and mark where the space from zero to 3/4 ends. Connect these tasks to students' prior experiences with measuring length and number lines. Provide real-world scenarios and context in which they measure the fraction. Have students measure how many wholes an object is and then explore how many halves, how many thirds, and how many fourths it is.

While working at the concrete level, students can also be shown how fractions are composed. Unit fractions, or equal-sized parts (e.g., 1/5), are combined to form 2/5. Understanding unit fractions are combined to form larger fractions forms the basis for more complex concepts related to computation. After students become proficient in making concrete models of fractions, show how this process can be represented using equations in which unit fractions are added. An example is shown in Figure 8–8.

After students use the fraction manipulatives to solve the fraction problems, instruction moves to the representational/semi-concrete level. Students need to be exposed to the area, set, and number line concrete models simultaneously before moving to the next phase with any of these models.

APPLICATION FOR BASIC FRACTIONAL UNDERSTANDING AT REPRESENTATIONAL/SEMI-CONCRETE LEVEL

Area

Instruction of the area model at the representational/semi-concrete phase involves students drawing fractions and partitioning them as well as interpreting fractions that are displayed. As discussed during the concrete phase, rectangles and circles are the most common visuals used for the area model, but both can cause struggles for students during the representational/semi-concrete phase. Therefore, it is important that students explore both visuals and that the teacher be explicit about the challenges each creates. It is difficult for students to partition a circle, when they move to the representation phase of CRA/CSA instruction. While it is easier for most students to recognize a fraction within a circle when the circle is already partitioned, teachers are discouraged from having students do the partitioning of circles.

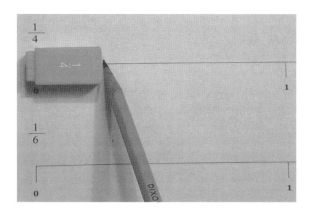
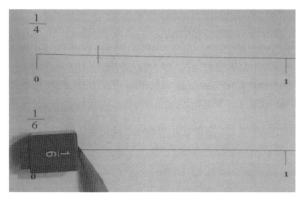

Figure 8–7. Marking Fractions on Number Lines.

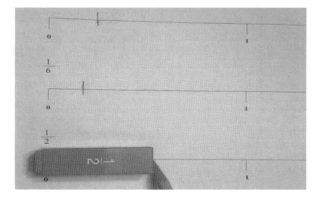
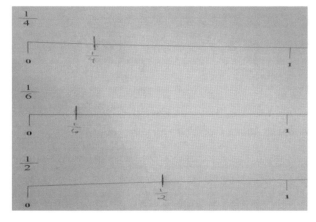

I am going to make three sixths and I want to show how I did it using an equation. So, my denominator is six and that means that there are six parts in the whole. Count the parts in the whole with me...1..2..3..4..5..6.

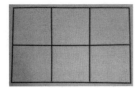

I am going to use three because the numerator is three. I am going to use three parts. First, I put one block. This is one sixth. If I want to show what I am doing using addition, I begin by writing the first part that I used... one sixth. What should I write first in my equation? Yes, one sixth.

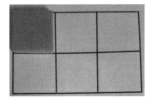

3		1							
—	=	—							
6		6							

Have I made three sixths yes? No, What am I doing with this next piece, am I combining it with the block that is already there? Yes. What operation do we use when we combine objects? Subtraction or Addition? Yes, addition. So, I write a plus sign.

$$\frac{3}{6} = \frac{1}{6} +$$

Figure 8–8. Explicit Instruction Modeling Addition of Unit Fractions. continues

Now I have two blocks, so I write another one sixth. The equation is one sixth plus one sixth

$$\frac{3}{6} = \frac{1}{6} + \frac{1}{6}$$

Am I finished? No, I need another sixth. What am I doing with this one sixth, adding it? Yes, so what symbol should I write in my equation? Yes, a plus sign

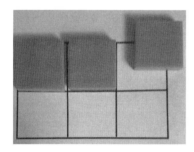

$$\frac{3}{6} = \frac{1}{6} + \frac{1}{6} +$$

I added another one sixth, so I can finish my addition equation. Three sixths equals one sixth plus one sixth plus one sixth.

$$\frac{3}{6} = \frac{1}{6} + \frac{1}{6} + \frac{1}{6}$$

Figure 8–8. continued

Rectangles are easier to partition, but the whole may be difficult to determine when students are creating rectangles to order, compare, and find equivalent fractions in the representation phase. In addition, rectangles may be partitioned horizontally, vertically, diagonally, or a combination of these. This will be discussed in that section of this chapter.

Moving to the representational/semi-concrete phase can sometimes involve showing a concrete example and a representational/semi-concrete example of similar shapes and asking students to make the connections. For instance, students may use pattern blocks to show fractions and then view drawings of pattern blocks depicting these same fractions. Helping students explicitly connect the concrete and representational/semi-concrete models is important. As students explore, the teacher will use mathematical language and write the equations for students to see how to write the symbolic form of what the illustration represents. For example, if a rectangle is divided into four pieces and three are shaded, the student will be encouraged to say, "3/4," and the teacher will write, "3/4" (Figure 8–9).

Keep in mind that it is important that examples begin with fractions less than one, but fractions with value greater than one need to be intertwined to avoid the misconceptions that all fractions are less than one. For example, have students find 3/4, 2/3, and then 4/3. This can be challenging but pushes them to examine what the entire fraction means.

Set

When moving to the representational/semi-concrete phase of the set model, students can often become distracted drawing the images with great detail, rather than simply using them as a visual to show what is being communicated. Just like the area model, it is important to help students make the connection between the concrete examples with the representational/semi-concrete examples. In addition, students need opportunities to draw images

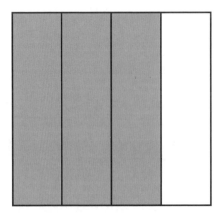

Figure 8–9. *Initial Representation-Level Instruction of the Area Model of Fractions.*

that represent the set model of a given fraction as well as being able to write the fraction of an image of the set model.

Begin representational/semi-concrete instruction by having a picture of 12 counters. Ask students to circle 1/2 of the 12 counters in the picture. After they circle six counters, ask them how many counters are circled and how many are not circled. Ask them what fraction of the 12 counters is circled (1/2) and also ask them to share the fraction of the counters that is not circled (1/2). The second problem should also show 12 images (counters, pictures of apples, etc.). Ask the student to show 1/3 of the images. Discuss the strategy the student used. Did he use the "one for me, one for person 2, and one for person 3" strategy or did he count the total and divide by 3? Based on preassessment of the student's multiplication and division skills, if possible, move him toward the latter strategy, as this is more efficient in the long run. Continue using an image of 12 items to show fourths, sixths, and twelfths. Then explore this with other total numbers of images. Then move to questions that have students draw examples of fraction sets using different numbers. These are often easiest when given a contextual problem. For example, I have 24 goldfish. Draw a picture showing how I could divide these into sixths. Then circle 2/6 or the goldfish. Initially you will ask students to do each step one at a time. As students become more comfortable, you will move to asking them to draw a picture to represent a fraction of a set such as 1/4" (Figure 8–10). It is important during this phase that students are sometimes asked to draw the set, sometimes asked to circle the predrawn set, and sometimes asked to write the fraction that the set represents.

Length Model

Representation of the length model allows students to connect and compare fractions. It also links their experiences of lining up manipulatives for counting

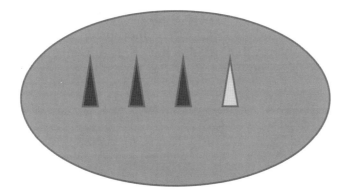

Figure 8–10. *Initial Representation-Level Instruction of the Set Model of Fractions.*

and using the number line to explore magnitude with whole numbers. When exploring single fractional numbers, drawing fractions using fraction bars is easier. Students can adjust the length of the whole they create to make equipartitioning easier for difficult amounts such as thirds, fifths, and so on. Helping students utilize their understanding of factors can help with some equipartitioning. For example, they can divide a rectangle into halves two times to make fourths. It is important that students think aloud how they are considering equipartitioning and that the instructor provide scaffolding advice to develop effective strategies. This will be extremely important when they move to comparisons and equivalencies.

Representational/semi-concrete instruction of the number line begins by having students draw a number line. Ensuring they are able to equipartition the zero, one, and two will be an important indication of their ease at equipartitioning fractions. If fine motor skills prevent the student from partitioning the number line, instruction can involve number lines that are already partitioned. However, eventually students will want to move toward partitioning the number line if at all possible.

Have students practice equipartitioning (dividing into equal segments) the number line into various pieces. For example, have students equipartition four segments between zero and one and four more segments between one and two. Ask students how many segments are in one whole (four) and how many segments are in two wholes (eight). Then repeat this with different numbers. Next move to having students demonstrate fractions on the number line. For example, show a picture of a number line divided into fifths and with an "x" marked on 3/5. Ask the student how many segments are used between zero and the "x." Review that the three of the five segments that make a whole are used before the "x." Do more fraction examples where the number line is already equi partitioned and the fraction is marked, then move to mixing examples that are less than one and examples that are more than one, such as 6/5. Students need experiences showing a fraction on the number line and then writing a fraction that is depicted on a number line (Figure 8–11).

In order for this to be relevant to students, it is important to also provide real-world scenarios and context where measuring to the fraction. Have students measure objects that are 1½ inches, 3/4 inch, and so on. As they measure these objects, point out that the fractions are equipartitioned on the ruler.

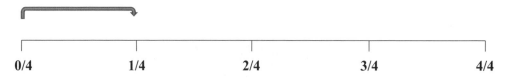

Figure 8–11. Initial Representation-Level Instruction of the Number Line Model of Fractions.

APPLICATION FOR BASIC FRACTIONAL UNDERSTANDING AT ABSTRACT LEVEL

The abstract phase of fraction understanding does not use the area, set, or number line model because students have internalized these models to create conceptual understanding of fractions. Asking students to describe fractions within contextual problems as well as discussing the meaning of various fractions is seen at the abstract phase. So asking questions such as:

1. If you had eight crayons, how many crayons would 3/4 of your total crayons be?
2. If your garden had five rows and two of the rows had cabbage planted, what fraction of the rows would have cabbage?

With abstract questions, it is important to ask follow-up questions to formatively assess student understanding and determine if any further remediation is needed. If students are still having difficulty without the visual cues, it is important to take a step back and encourage them to use these cues to make sense of the problem. The sense-making should be prioritized above the abstract understanding.

COMPARING, ORDERING, AND IDENTIFYING EQUIVALENT FRACTIONS

It is important that students develop multiple ways of comparing and ordering strategies in order to select the most efficient method and to have a holistic understanding of the relationship between fractions (Petit et al., 2016). Fractions comparisons can involve (a) different denominators and same numerators, (b) different numerators and same denominators, and finally (c) different numerators and different denominators (Behr, Wachsmuth, Post, & Lesh, 1984).

Manipulatives and visual representations can be useful as students develop skills to compare and order fractions. This is why CRA/CSA can be a beneficial instructional model to support student growth. Students first use reasoning related to unit fractions to help when comparing and ordering fractions. This is very effective when comparing fractions with different numerators and the same denominator (such as 2/5 and 3/5). Students also use benchmarks in comparing and ordering fractions. Benchmarks, also known as referent points, can also be helpful. For example, 2/5 is less than 1/2 and 4/6 is more than 1/2. Therefore, one can conclude that 2/5 is less than 4/6. Before teaching students other ways to compare fractions, they must conceptually understand fractions, their values based on examination of numerators and denominators, and basic operations with fractions. With these conceptions mastered, students can compare fractions by finding common denominators through multiplication and division. Comparing fractional values requires a strong understanding of equipartitioning as well as equivalence. It is essential that students have a firm understanding of the meaning of a fraction before they are asked to find equivalent fractions (Petit et al., 2016).

APPLICATION FOR COMPARING, ORDERING, AND EQUIVALENT FRACTION AT CONCRETE LEVEL

When providing intervention to students about comparing, ordering, and finding equivalent fractions, it is essential to provide multiple concrete models to help students generalize, rather than only learning what one model has to offer. Comparing and ordering fractions using different models is as important as using the CRA/CSA model for most students. Seeing how the meaning of the fractions can be interpreted and that the comparisons have the same results whether the fraction is represented in the set, number line, or area model is important. However, when exploring and linking the models, connecting the concrete, representational/semi-concrete and abstract representations can also be helpful. The concrete phase begins by comparing various fractions. Once students are comfortable comparing two fractions, teachers should explicitly help students use these comparisons to order multiple fractions. Equivalent fractions at the concrete phase are similar to other comparisons. "Equivalent fractions are fractions that represent equal value; they are numerals that name the same fractional number" (Chapin & Johnson, 2006, p. 114). They represent the same point on a number line or the same portion of a whole. Teachers can write the equivalent fractions as they are found during the concrete stage of comparing fractions and display these around the room. When students move to the representational/semi-concrete and abstract phases, these findings can be explored for patterns that lead to the algorithm that is often used to find equivalent fractions. This will be described in the abstract phase of instruction.

Area

While students can visually compare circle, rectangular, or square fractions manipulatives when the denominator is the same, the numerator is the same, or when there is a vast difference, other comparisons are made by laying the pieces on top of each other to see which is greater or if they are equivalent. This practice is exploratory and shows students that all fractions can be compared. In addition, using concrete models that students are familiar with from other content, such using the hexagon from pattern blocks as the whole, helps students understand that the whole does not just come in the form of a circle or a rectangle.

When a fraction has the same denominator (such as 2/5 and 4/5), students are able to visually see how many pieces of the whole are being used in the fraction, so the more pieces that are used, the larger the value (Figure 8–12). Concrete manipulatives can help make sense of this concept. When fractions with the same numerator but different denominators are used (such as 1/3 and 1/5), students are able to see how the larger denominators require smaller pieces (fractions) to make the whole. Therefore, they are able to generalize that when comparing fractions with the same numerator but different denominators, the largest fraction is the one with the smallest denominator. When students are asked to compare fractions with different

	Same Numerator/ Different Denominator	Same Denominator/ Different Numerator	Different Numerator & Denominator (equivalent)
Area-circle	1/2 1/3 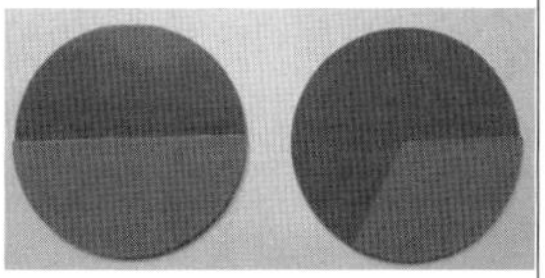	2/4 3/4 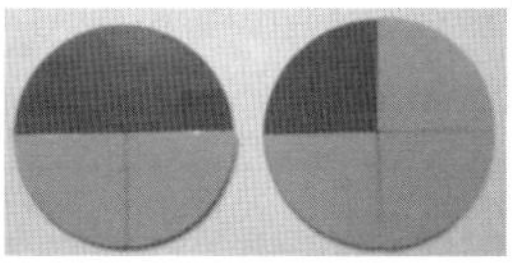	1/2 2/4 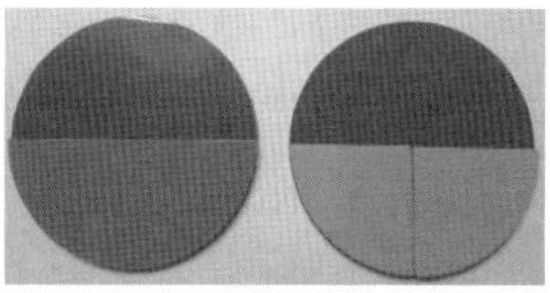
Area-rectangle	1/2 1/3 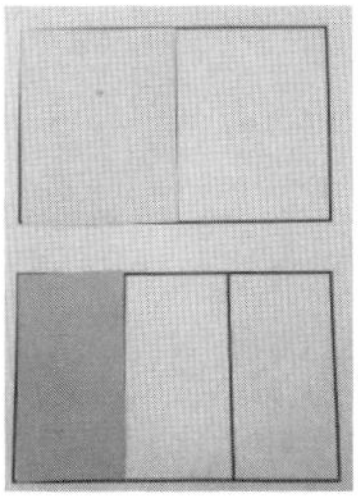	1/3 2/3 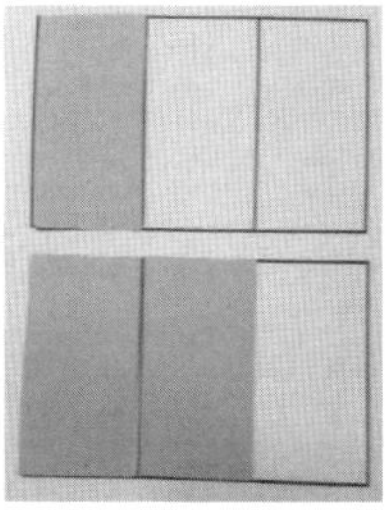	1/2 2/4 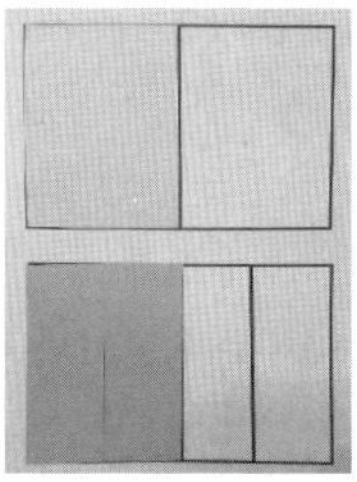

Figure 8–12. Using the Area Model for Comparison at Concrete Level. continues

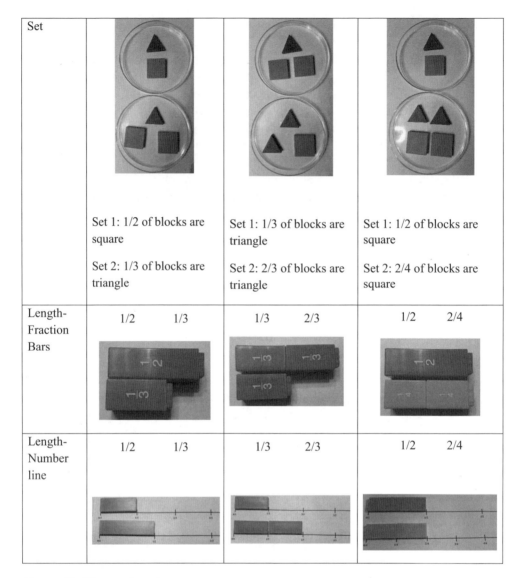

Figure 8–12. continued

numerators and denominators, they are able to visually see the difference using concrete objects, but generalizations and patterns are not obvious; therefore, more time should be spent discovering patterns and generalizing fractions beyond simply finding that one can compare fractions using concrete manipulatives.

In order to explore equivalency at the concrete level, circle fraction pieces can be useful (but pattern blocks and other manipulatives should be used as well). To lay the foundation for equivalency, when students find fractions that are equivalent, write the fractions out. If students compare 1/2, 2/4, 3/6, 4/8, and 5/10, they may begin to see a pattern that leads to understanding of equivalency. It is critical that the instructor know his or her students and determine if the student has a foundational understanding of multiplication that would enable them to determine the pattern with this support.

Set

This becomes much more complex for comparing fractions. Rather than thinking about area or distance, students need to be able to divide multiple objects into equal groups. Teachers can begin by asking students whether they think 2/3 of 12 counters or 2/4 of 12 counters would have more counters. After students predict and explain their reasoning, teachers can lead students through an exploration to test their predictions and explore the concrete representation of this problem. Teachers should try to let students take the lead in each of these steps as much as possible, but the directions are very detailed to help where needed. Using two sets of 12 counters (one set can be blue and one set can be yellow so that students do not confuse the counters), have students divide the blue counters into thirds and then count how many counters make up 2/3 of the 12 (8 counters). Then have students divide the yellow counters into fourths and count how many counters make up 2/4 of the 12 (6 counters). This helps students concretely see how 2/3 would be more than 2/4 when using the same whole number in a set. This comparison of same numerator but different denominators helps students see that the small denominators would mean the total number of items in the selected group (numerator) would be bigger than the larger denominators. When sets are divided into the same number of groups (denominators) but a different of these groups are selected (numerators) are compared, the fraction representing the largest number of groups selected (the one with the largest numerator) will have the most items. For example, if the 12 blue counters are divided into three equal groups (4 in each group) and the 12 yellow counters are divided into three equal groups (with 4 in each group), these both represent the division of the counters into thirds. Asking students to count only 1/3 of the blue counters and 2/3 of the yellow counters and then compare which have more counters helps the students visualize this complex topic. As an intervention teacher, it is important to ask students to predict before actually walking through the comparison. Also, allowing students to take the lead and share reasoning for each step is important. The description above is intended to help the interventionist see the important steps the teacher will scaffold for the student and ensure comparisons of fractions in sets are understood. After multiple experiences where students have begun to see the patterns and make generalizations, then teachers can have students compare fractions with different numerators and denominators, such as 2/3 and 3/4. However, these comparisons must only be made when they have the same number of counters composing the whole (such as the 12 counters for all fractions listed above). This is an important concept that students and teachers must understand before proceeding. Repeat these comparisons in different ways.

Length

Often the concrete representation of fraction bars is useful in comparing and ordering fractions. In fact, many classrooms make their own fraction bars from construction paper, which also helps with equipartitioning. Then these physical manipulatives can easily be compared. It is important to ensure students understand that to compare bars, they must be lined up such that comparisons can be seen visually. For example, if measuring horizontally, both fraction bars being compared need to be lined up starting at the same horizontal point with one bar being above the other (see Figure 8–9). If a fraction with more than one unit such as 2/3 or 3/4 is being compared, it is important that each unit be touching the end of the previous unit. For example, 1/3 touches the next 1/3 without space between these, which might distort the actual length. This is similar to the struggle students sometimes have with measuring objects and not beginning at the zero on the ruler or spacing between objects and accidentally measuring this space. While most students will naturally compare this way, it is important that the instructor watch for these errors that could impact the visual comparisons.

The fraction bars show equivalency, because students can compare all bars together and see where fractions line up and display equivalency. As students share equivalent fractions, teachers can write the fractions to find patterns, similar to what is suggested with the area model. These comparisons and discussion about patterns at various points of intervention can help lay seeds toward understanding the abstract concept of multiples, which is a common strategy when moving to the abstract phase.

Number lines provide another visual for comparing fractions. While students may benefit from physically partitioning and marking fractions on the number line, the representational/semi-concrete phase is similar to the concrete phase when using this model because the number line is a pictorial representation. During the concrete phase, students physically demonstrate the space represented on the number line, but they mark space by drawing on a number line just as in the representational/semi-concrete phase. Therefore, less time may be needed.

While most of the discussion during this concrete phase involves comparing fractions with the same whole, it is also important to discuss questions such as whether 1/2 is always greater than 1/3 within the context of different wholes. It is important that students do not overgeneralize conceptions of fraction comparison to the point that they believe that one-half is always larger than one-third despite the whole objects. For example, one half of a cupcake is not more dessert than one third of a cake because the whole objects are different. The same is true of sets; one half of a set of four is not more than one third of a set of 24. Students must understand that comparisons can only be made between fractions that are formed from wholes that are the same in size and magnitude. An example of instruction is provided in Figure 8–13.

We are going to consider this question: **Is one half always bigger than one third?**

Look at these two desserts, a cupcake and cake. How are they different? Are they the same size? When would you have more dessert, with a cupcake or a cake?

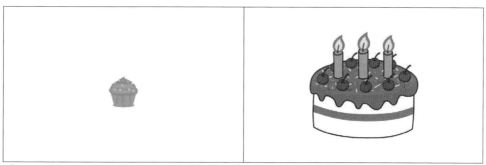

If we said that one half is always more than one third, would one half of this cupcake be more than one third of this cake?

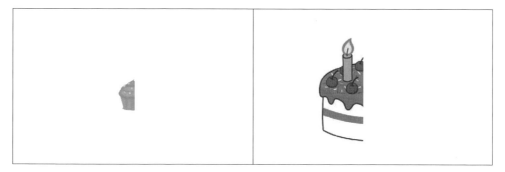

So, we need to think about the wholes. Should they be the same or different in size when we compare fractions? The same.

Figure 8–13. Demonstrating Importance of the Whole When Comparing Fractions. continues

*So, the answer to our question about one half always being bigger than one third should include something about the whole. When the wholes are the same, one half will be bigger than one third. So, one half of the cake will always be bigger than one third **of the same cake.***

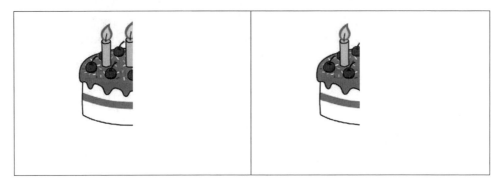

We are going to consider this question: **Is one half always bigger than one third?**

Look at these sets. How are they different? Are they the same size? How many triangles? 12 *How many squares?* 18

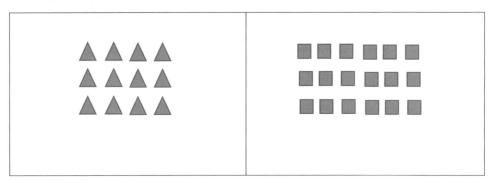

If we said that one half is always more than one third, would one half of the triangles be more than one third of the squares? Let's find out. How many parts in the whole for ½? Two. *We make two equal parts. How many parts do we use?* One. *So, we circle one of the parts. How many did we circle?* Six

Figure 8–13. continues

How many parts in the whole for 1/3? Three. *We make three equal parts.* How many parts do we use? One. *So, we circle one of the parts.* How many did we circle? Six *The whole sets are not the same, so one half is NOT bigger than one third.*

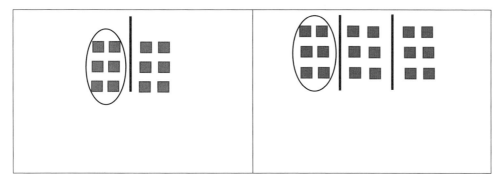

So, we need to think about the wholes. Should the whole set be the same or different in size when we compare fractions? The same.

*So, the answer to our question about one half always being bigger than one third should include something about the whole. When the wholes are the same, one half will be bigger than one third. So, one half of 18 will always be bigger than one third **a set of 18.** One half of 18 is 9 and one third of 18 is 6.*

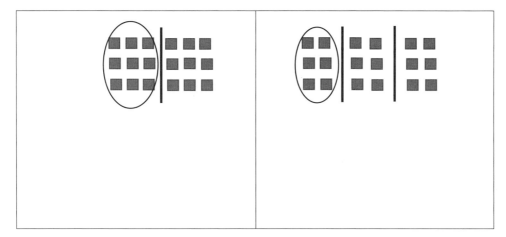

Figure 8–13. continued

APPLICATION FOR COMPARING, ORDERING, AND EQUIVALENT FRACTION AT REPRESENTATIONAL/SEMI-CONCRETE LEVEL

Area

The circular pieces in the representation phase should only be compared when the pieces are already equipartitioned, because this is extremely challenging with circles. The rectangle fraction pieces are easier to partition, but the whole may be difficult to determine, so initially it can be helpful to have the whole already drawn when comparing, ordering, and exploring equivalency during the representation phase of CRA/CSA instruction. Before beginning with the pictures, it can be useful to explore virtual manipulatives through websites or applications that allow the students to manipulate, equipartition, and create their own fractions through technology. This isn't the same as being able to physically manipulate the manipulatives as students did in the concrete phase, but it is a way to transition students into viewing two-dimensional representations.

Using the area model can help when comparing like denominators or like numerators. Students are able to see that with like denominators, the larger the numerator, the more pieces (which are the same size), so the larger the amount. For example, with 4/12 compared to 6/12, students are able to see that 6/12 has more pieces and is therefore a larger number than 4/12. This is an example of like denominators. An example of like numerators would be comparing 2/3 to 2/5. Students are able to see that as the denominator becomes larger, the size of each piece becomes smaller. Although pictures cannot be physically moved to lay on top of each other, these same principles can be explored with pictures. It is important to also compare fractions using various pictorial representations. So comparing 1/2 to 2/3 using fraction circles, rectangles, and pattern blocks allows students to generalize that if the whole is the same, 2/3 is always larger than 1/2.

In the representational/semi-concrete phase, the area model can help with comparing and ordering, unless they are close to equivalent. The visuals are two-dimensional, so it is difficult to determine fractions that are close to equivalent and fractions that are actually equivalent. For example, a pictorial representation of 11/12 and 23/24 will look very similar. However, when they look close, instructors can push students to use other strategies. For example, at this point students should understand that 1/12 is larger than 1/24, so that means 11/12 is further from one than 23/24 is. This can be a confusing concept when told, but the visual and questioning to scaffold this can create a sense that fractions should make sense and "we don't have to just guess" when the fractions look close. Figure 8–14 shows a representational/semi-concrete model for equivalent fractions.

The rectangular area model is useful for comparing fractions and finding equivalent fractions. When comparing fractions and trying to determine

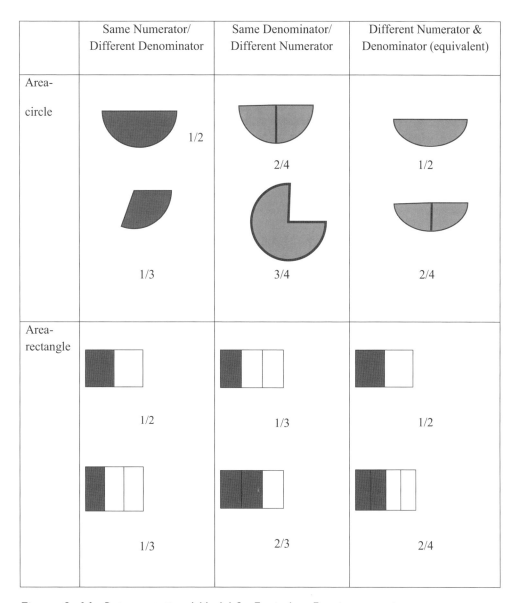

Figure 8–14. Representational Model for Equivalent Fractions. continues

which fraction is greater, one can draw two rectangles that are the same size. For example, if students are asked to compare 2/3 and 3/5, the student will draw 2/3 on the first fraction and the thirds are created using horizontal lines. On the second rectangle, they will illustrate 3/5 using vertical lines. To compare these two, we can find common denominators by marking the vertical lines on the 2/3 and the horizontal lines on the 3/5 to make the first rectangle illustrate 10/15 and the second rectangle illustrate 9/15 (Figure 8–15). Using different colors when creating thirds and fifths can be useful in helping

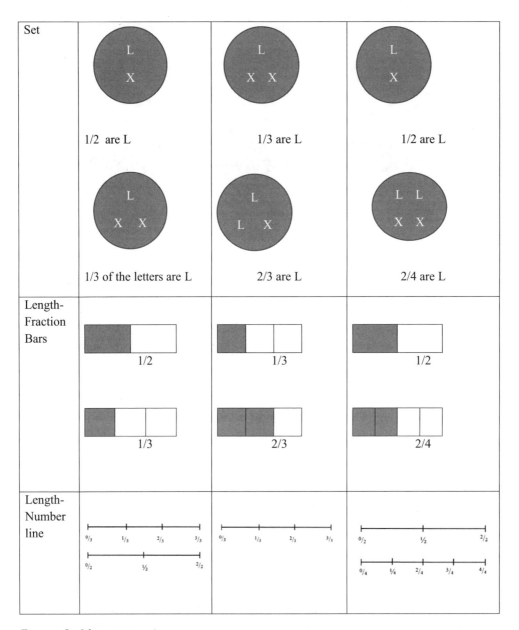

Figure 8–14. continued

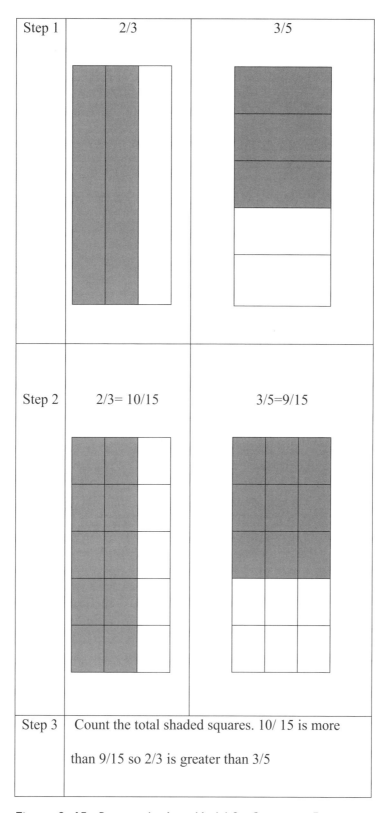

Figure 8–15. Rectangular Area Model for Comparing Fractions.

students understand what happened. This illustrates that the actual fractional amount didn't change, but the number of pieces has changed.

When students are given a fraction such as 2/3 and are asked to find an equivalent fraction, they can follow a similar process to the comparison model illustrated in Figure 8–15. Students could create 2/3 using horizontal lines and then use vertical lines to change the fraction without changing the amount represented on the representation. So 2/3 could become 4/6 or 6/9. It is important to discuss with students how the amount didn't change, the whole didn't change, and how the pieces of the whole are still equi-partitioned. These are misconceptions students often make when exploring equivalency of fractions.

Set

If the instructor is challenged with not enough intervention time to completely support the student's attainment of all skills, the set model would be the suggested model to skip at this phase. It is the least utilized one. However, if there is time, it does provide an opportunity to see how fractional comparisons of the same whole result in the same ordering despite the model used.

With the representational/semi-concrete phase, students can be asked to explore problems similar to those described in the concrete phase. The difference being that the fraction representations cannot be moved because pictures are used. Therefore, having two copies of the drawing of sets is needed. For example, having two drawings of 12 circles allows students to then divide the first drawing to show 2/3 and the second drawing to show 2/4 and then compare these sets. It is important that both sets have the same number. It is also critical if students are drawing the sets that the instructor remind the students the drawings should be simple so they can focus their time on the mathematics, rather than spending the time drawing detailed pictures. One helpful bridge between concrete and representational/semi-concrete is the use of virtual manipulatives. Technology enables users to manipulate two-dimensional images of fractions and transition from the concrete manipulatives they can physically touch to the representational/semi-concrete pictorial representation of fractions that cannot be moved to make direct comparisons.

Length

When comparing fractions using the length model at the representational/semi-concrete phase, it is important to begin with pictures of fraction bars that are placed below each other as they were during the concrete phase of instruction. Also having fraction bars available to help students see the connection between the pictures and concrete manipulatives is important. One bridge can be virtual manipulatives. While these are not concrete, they are able to be manipulated more than semi-concrete pictorial representations. Students are able to partition using many virtual manipulatives and also move the fractions to directly compare various representations.

Comparing fractions with the same numerator or the same denominator using the representations of the bar model allows students to review the meaning of fractions and draw conclusions about what happens as the numerator and/or denominator changes, such as described in previous sections. Pictures of fraction bars also help students begin to draw conclusions using benchmarks. For example, asking students if the fractions are closer to zero, one half, or one helps students begin to develop flexibility in their thinking. When students recognize that 1/10 is close to zero and 3/5 is close to 1/2, they can compare and order these two fractions. This will also help as they move to computation with fractions.

Students are able to equipartition pictures of bar fractions to change the fractions. For example, 1/2 can be split in half to become 2/4. This partitioning at the representational/semi-concrete phase lets students see that the amount doesn't change, but that the number of pieces in the whole and the number of pieces being used are changed at a proportional amount. Teachers can write the original fraction (such as 1/2) and then ask students to explore how both the number of pieces in the whole and the number of pieces being used were multiplied (such as multiplying by 2) to create an equivalent fraction (2/4). Then the teacher can ask the student to make predictions of another thing they could do to create an equivalent fraction (such as divide in half again, which is the same as multiplying the numerator and denominator by 2 to make 4/8). These discussions will be based on student understanding, but it is important that students can change the fractions to create new equivalent fractions and that they begin to see the role of factors and multiples in this change in order to move to the abstract phase.

Number line models are extremely useful when comparing, ordering, and finding equivalency. The number line is a visual for exploring magnitude. One useful tool is the benchmarks of zero, one half, and one that allow students to begin to gain concepts of fractions relative to other numbers and points on the number line. This concept of relationships between magnitude in numbers is explored in great depth with whole numbers but is often neglected when exploring fractions. The number line is often used in exploring relationships with whole numbers when students are first learning about single-digit numbers. Fractions add a layer of complexity, because the number of partitions depends upon the fractions being explored (such as fourths that have four partitions and ninths that have nine partitions). To begin, teachers should ask students to label the number line with zero and one and ask students to place 1/2 where they think this goes. Then students should partition number lines to show various unit fractions that add to one, such as partitioning and labeling 1/4, 2/4, 3/4, and 4/4. Progressing to ordering and comparing, students need experiences estimating where various fractions belong on the number line in relation to the three points already labeled. For example, students should share whether 2/5 is closer to zero, one half, or one and explain why. If needed, the teacher can have the student partition all fifths to one or 5/5 as they did previously. Mastery of relating fractions to the benchmarks of zero, one half, and one is essential before students can move to comparing other fractions.

Once students can compare benchmark numbers to fractions, then the instructor can move toward comparing and ordering two different fractions. For example, comparing 1/5 to 5/6 allows students to recognize that 1/5 is closer to zero and 5/6 is closer to one, so 5/6 is larger than 1/5. Beginning with fractions that are easy to compare and visualize on the number line allows students to develop flexibility in thinking. Then, more difficult fractions are explored at the representational/semi-concrete level of the number line. For example, 3/4 and 5/6 can be compared in multiple ways using the number line. Teachers should begin by asking students what they think they can do to solve it. Providing two number lines parallel to each other, with one equipartitioned and labeled with fourths and the other number line labeled with sixths, could scaffold the learning. Students can mark an "X" on the spot for 3/4 on the number line partitioned into fourths and an "X" on the spot representing 5/6 on the number line partitioned into sixths. It is essential that the zero and one on the number lines are at the same spot on both number lines and that the unit fractions are partitioned equally. Ask students to compare and discuss what they see. Some will be able to determine which is larger; sometimes it is difficult to see, but they can compare how far from one the two fractions are (e.g., 3/4 is further from one than 5/6 is). Students' understanding of denominators is important to understanding this example. Partitioning a number line into four parts will result in larger parts as compared to partitioning the same line into five parts. Both 3/4 and 5/6 are one unit (1/4 and 1/6) away from one (4/4 or 6/6). For the fraction 5/6, the space to one is smaller than that of 3/4 because each unit is smaller. Therefore, 5/6 is closer to one than 3/4 and, therefore, greater in magnitude.

Multiple experiences with this are important. As students become more comfortable with the number line, you can encourage them to move to one number line and they can partition in different colors or have one fraction partitioned on the top of the number line and one partitioned on the bottom to compare these (Figure 8–16).

In addition to comparing and ordering fractions, use the number line to find equivalent fractions. Have students begin by drawing two number lines of the same lengths parallel to each other that have zero, one half, and one labeled. Partition the first number line to illustrate the fraction given such as 2/3. They may use the second number line and mark 2/3, but discuss how to equipartition to make a different fraction that represents the same amount. For example, add a hash mark halfway between each third to create 4/6 or add a hash mark dividing each third into thirds and make 6/9. An important

Comparing one half and three fourths

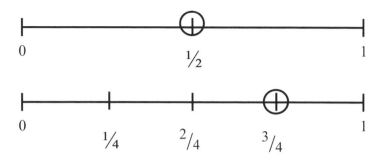

Comparing three sixths and one half

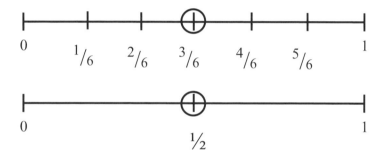

Comparing three fourths and five sixths

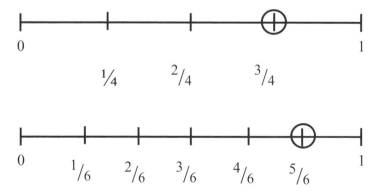

Figure 8–16. *Comparing Fractions Using the Number Line.*

part of this interaction is that the point on the number line doesn't change and that each unit fraction must be equally partitioned to create the new number. Discussion of multiples and factors can be useful to help students realize these are tools that help when you move to the abstract phase. For example, when comparing number lines with thirds and sixths, show that it takes two sixths for each one third. This forms the basis for the abstract approach to making equivalent fractions with multiplication and division.

APPLICATION FOR COMPARING, ORDERING, AND EQUIVALENT FRACTION AT ABSTRACT LEVEL

Students often think that the value of a number changes when the numerator and denominator are multiplied or divided by the same number (McNamara & Shaughnessy, 2015). Because students have had experiences seeing equivalency using the concrete and representational/semi-concrete fractions, students have experienced how the fractional notation changes, but the value of the number doesn't change. It is important to revisit this idea through each phase of CRA/CSA and to bring it back up as students move to abstract. Students may be able to compare and order most fractions based on their number sense that has been developed throughout the concrete and representational/semi-concrete phases. They have seen fractions compared using the set, area, and length models. Now students move to comparing without these visuals. One way to compare when the answer isn't obvious is to find a common denominator. At the concrete and representational/semi-concrete phase, the instructor has discussed patterns and helped students notice that there are multiples in finding equivalent fractions. During these phases, the teacher asks questions to help students realize that multiplication or division of both the numerator and denominator leads to the equivalent fraction without changing the value. For example, visual input via area models or number line showed that it takes three ninths for each one third. At the abstract phase, students learn to compute this, but remind them of the other phases so they recognize that while the numbers have changed, the value has not. Students at this point of instruction understand fractions can equal one. Students multiply or divide both the numerator and denominator by the same number or a fraction equal to one. This creates an equivalent fraction; the value is unchanged, just like multiplying or dividing any whole number does not change the original value (e.g., $3 \times 1 = 3$ and $1/3 \times 2/2 = 2/6$). Another example is, $1/2 = 2/4 = 4/8$. One half is multiplied by $2/2$ to change the fraction notation to $2/4$. Two fourths is multiplied by $2/2$ to change the fraction to $4/8$. Two halves (or any time one multiplies or divides by a number with the same numerator and denominator) is equal to one. Students should be aware of the identity property, which they learned when exploring whole-number multiplication. The identity property states that any number multiplied or divided by one equals the original number. So, although $1/2$ and $2/4$ look different, the value remains the same. It may be helpful to begin instruction at the abstract level with pictures (e.g., two number lines representing two equivalent fractions) so that students can understand why and how the abstract procedure is used. The authors of this book implore teachers to ensure that conceptual understanding is firm prior to teaching students procedures at the abstract level. In order for students to use their understanding of fractional numbers and be successful in later mathematics learning, they must understand how and why shortened procedures are used. Simply memorizing procedures will lead to confusion, error patterns due to poor memory, or mixing procedures to form error patterns.

CHAPTER SUMMARY

This chapter provided examples of interventions using the CRA/CSA sequence to support students struggling with concepts related to fractional understanding. Sufficient development of fractional understanding involves interpretation and manipulation of fraction numbers using multiple models such as area, set, and length. The CRA/CSA sequence should be used to assist students in understanding each of these models, and this chapter provided instructional examples and suggestions. The beginning concepts of fractions learned through third grade were presented and more complex concepts related to operations and decimals are addressed in the next chapter. Prior to engaging in more complex tasks related to fractions, it is critical that students have firm conceptual understanding of fractions, well beyond simply recognizing the shaded portion of a shape. Students who struggle in mathematics may come to fraction intervention instruction with weak number sense and reasoning, skills needed in order to advance to more complex understanding of numbers, including fraction numbers. As demonstrated in this chapter, provide students with experiences that encourage exploration and reasoning along with guidance in that exploration as well as explicit instruction.

APPLICATION QUESTIONS

1. How are whole numbers related to fractional concepts?

2. Using any approach, how would instruction in beginning fractional concepts be taught at each of the levels of the CRA/CSA sequence?

3. What is the rationale for using each of the following models when teaching fractions: set, region, and number line? How does each of these models benefit students?

4. Provide reasons why one might choose each of the following approaches to comparing fractions: unit fractions, benchmarks, and finding common denominators.

REFERENCES

Behr, M., Wachsmuth, I., Post, T., & Lesh, R. (1984). Order and equivalence of rational numbers: A clinical teaching experiment. *Journal for Research in Mathematics Education, 15,* 323–341.

Bezuk, N. S., & Bieck, M. (1993). Current research in rational numbers and common fractions: Summary and implications for teachers. In D. T. Owens (Ed.), *Ideas for the classroom: Middle grades mathematics* (pp. 118–136). Reston, VA: National Council of Teachers of Mathematics.

Butler, F. M., Miller, S. P., Crehan, K., Babbitt, B., & Pierce, P. (2003). Fraction instruction for students with disabilities: Comparing two teaching sequences. *Learning Disabilities Research & Practice, 18*(2), 99–111.

Chapin, S. H., & Johnson, A. (2006). *Math matters: Understanding the math you teach, Grades K–8.* (2nd ed.). Sausalito, CA: Math Solutions.

Confrey, J., Maloney, A. P., Wilson, P. H., & Nguyen, K. H. (2010, April). *Understanding over time: The cognitive underpinnings of learning trajectories.* Paper presented at the annual meeting of the American Education Research Association, Denver, CO.

Confrey, J., Nguyen, K., Lee, K., Panorkou, N., Corley, A., & Maloney, A. (2010). *Turn-on Common Core Math: Learning trajectories for the Common Core State Standards for Mathematics—Overview of fraction trajectory.* Retrieved from http://www.turnonccmath.net

Cramer, K., & Whitney, S. (2010). Learning rational number concepts and skills in elementary school classrooms. In D. V. Lambdin & F. K. Lester Jr. (Eds.), *Teaching and learning mathematics: Translating research for elementary school teachers* (pp. 11–22). Reston, VA: NCTM.

Empson, S. B., & Levi, L. (2011). *Extending children's mathematics: Fractions and decimals.* Portsmith, NH: Heinemann.

Lesh, R., Landau, M., & Hamilton, E. (1983). Conceptual models in applied mathematical problem solving research. In R. Lesh & M. Landau (Eds.), *Acquisition of mathematics concepts & processes* (pp. 263–343). New York, NY: Academic Press.

McNamara, J., & Shaughnessy, M. M. (2015). *Beyond pizzas and pies* (2nd ed.). Sausalito, CA: Math Solutions.

Petit, M. M., Laird, R. E., Marsden, E. L., & Ebby, C. B. (2016). *A focus on fractions: Bringing research to the classroom* (2nd ed.). New York, NY: Routledge.

Saxe, G. B., Gearhart, M., & Seltzer, M. (1999). Relations between classroom practices and student learning in the domain of fractions. *Cognition and Instruction, 17,* 1–24.

Saxe, G. B., Shaughnessey, M., Shannon, A., Langer-Osuna, J., Chinn, R., & Gearhart, M. (2007). Learning about fractions as points on a number line. In W. G. Martin, M. E. Strutchens, & P. C. Eliot (Eds.), *The learning of mathematics 2007 yearbook* (pp. 221–236). Reston, VA: National Council of Teachers of Mathematics.

Siebert, D., & Gaskin, N. (2006). Creating, naming, and justifying fractions. *Teaching Children Mathematics, 12*(8), 394–400.

Sowder, J., & Wearne, D. (2006). What do we know about eighth-grade achievement? *Mathematics Teaching in the Middle School, 11*(6), 285–293.

Tirosh, D., Fischbein, E., Graeber, A., & Wilson, J. (1998). *Prospective elementary teachers' conceptions of rational numbers.* Retrieved from: http://jwilson.coe.uga.edu/Texts.Folder/Tirosh/Pros.El.Tchrs.html

Witzel, B. S., & Allsopp, D. (2007). Dynamic concrete instruction in an inclusive classroom. *Mathematics Teaching in Middle School, 14*, 244–248.

Witzel, B. S., Mercer, C. D., & Miller, M. D. (2003). Teaching algebra to students with learning difficulties: An investigation of an explicit instruction model. *Learning Disabilities Research and Practice, 18*, 121–131.

CHAPTER 9

Operations With Fractions Using the Concrete-Representational/ Semi-Concrete–Abstract Sequence

OVERVIEW

This chapter builds on the previous chapter. This chapter illustrates how computation of fractions connects to understanding fractional meanings, comparisons, and equivalencies. It also stresses the importance of connecting computation of fractions with computation of whole numbers. Too often, procedures are taught without conceptual understanding or contextual connection for fractional computation. This chapter addresses instruction emphasizing conceptual understanding through the concrete-representational/semi-concrete–abstract (CRA/CSA) sequence that allows students to form deeper understanding of these processes. The CRA/CSA sequence will be discussed with respect to addition, subtraction, multiplication, and division of fraction numbers. Fractions with common denominators and unlike denominators will be included in discussions of each operation. The concepts are presented with examples that encourage exploration as well as explicit instruction if students need more intensive experiences.

SEQUENCE OF INSTRUCTION FOR OPERATIONS WITH FRACTIONS

Operations that include fraction numbers are first presented using unit fractions, a topic discussed in the previous topic. Students learn that fraction numbers comprise units that are combined (e.g., 3/4 = 1/4 + 1/4 + 1/4). From there, students add and subtract fractions that have like denominators. After students add and subtract fractions with the same denominator, they learn how to multiply fractions. It is important that students can demonstrate conceptual understanding of multiplying by fractional numbers, rather than just learning the procedure (top times top and bottom times bottom). After students learn multiplication, concepts related to equivalence are expanded upon. Students learn about equivalence using concrete objects and pictorial representations prior to using them to complete operations. However, the process of changing a fraction can occur once students understand multiplication. Once students can change fractions into equivalent fractions, instruction in addition and subtraction of fractions with unlike denominators is possible. Next students learn about division of fractions. Similar to multiplication instruction, it is essential that students can demonstrate their conceptual understanding of division of fractions rather than just memorize a procedure (e.g., flip second fraction and multiply). Fraction concepts carry into advanced mathematics, and simply memorizing procedures will not allow for sufficient mathematical understanding to be successful in prealgebra, algebra, and beyond.

DESCRIPTION OF COMPUTATION OF FRACTIONS AND PREREQUISITE SKILLS

In order to begin computation of fractions, students need a deep understanding of whole-number computation and of fractional concepts. Exploring how a problem can be solved in multiple ways helps develop flexibility in thinking and conceptual understanding (Blöte, Van der Burg, & Klein, 2001; Rittle-Johnson & Star, 2007; Star & Seifert, 2006). Often students have been introduced to incorrect generalizations during whole-number computation, such as that the answer is always bigger in multiplication and that addition is easier than multiplication. While these statements may be true of whole-number computation, they do not necessarily ring true as students progress into rational numbers and integers. So when the "rules" seem to switch or not apply, students lose sense-making of how computational meaning is the same for these. Instead, they often learn the procedures at the expense of conceptual understanding and connections of meanings across problem types. Therefore, it is essential that teachers in earlier graders avoid teaching these rules that expire. When teaching computation of fractions, teachers need to acknowledge some of the assumptions students may have made with whole-number computation and explore why these do or don't work with rational numbers. Research indicates that students often make computational errors with fractions, because they are misapplying their understandings of whole numbers (Byrnes & Wasik, 1991; Kelly, Gersten, & Carnine, 1990; Ni & Zhou, 2005).

It is important that students have strong conceptual understanding of fractional meanings before beginning computation. Students should understand that fractional parts are equipartitioned. In other words, pieces are equal in the area model, sets are equal in the set model, and on a number line, there are equal spaces between unit fractions. This essential understanding is important when comparing many fractions that had different numerators and denominators (as described in the previous chapter). It will also be important when adding or subtracting with these different types of fractions. As students begin to add and subtract, they will need to retain a solid ability to identify the whole when working with multiple fractions. Therefore, much work about identifying and exploring the whole needs to occur prior to introducing computation of fractions. Also, students need experiences with mixed and improper fractions.

They also need to have some understanding of comparing, estimating, and ordering fractions. For example, students should recognize that 9/10 is closer to 1 than 1/2 or 0. This relational understanding of fractions allows them to use relational thinking strategies in computation (Empson & Levi, 2011). Students should be able to recognize that 4/5 is four units of 1/5 (1/5 +1/5 + 1/5 + 1/5). Extensive work with the number line builds a foundational sense of magnitude of various fractions that will be important in computation. Equivalency is a foundational skill when students are adding and subtracting fractions with different denominators. Conceptualizing that the amount or value of the fraction doesn't change, but the number of parts/pieces in the whole is all that has changed is a difficult concept, but very important. Ultimately, instruction on computation shouldn't begin until students are comfortable talking about fractions, comparing fractions, and expressing them both abstractly and visually.

When computation has been introduced to the whole class but some students need further intervention support, CRA/CSA is a model that can be implemented. It can help students conceptually understand what is occurring in the abstract computation. It allows students to see how computation of fractions follows the same basic principles as computation of whole numbers. Students need experiences with manipulatives and visual representations to make sense of problems before learning abstract algorithms (Kamii & Warrington, 1999). Concrete and representational/semi-concrete modeling also helps students see how and why some of the misconceptions that they built while exploring whole numbers are false, such as adding is easier than multiplying. Concrete and representational/semi-concrete experiences will help students understand the abstract procedures and equations they will use in later grades.

ADDITION AND SUBTRACTION OF FRACTIONS AT THE CONCRETE LEVEL

When providing intervention support for addition (and later subtraction) of fractions, it is important that students are able to see multiple models. The area and length models (see previous chapter) are both excellent models to build a holistic understanding of this process. The CRA/CSA sequence can be applied to either model.

Area

To begin, teachers should ensure students are able to use models to add fractions with like denominators. For example, using rectangle-shaped fractions, have students use the fraction pieces that are divided into thirds. Ask students how many thirds make one whole. As they share, write out 1/3 + 1/3 + 1/3 = 1 whole. Teachers should have students put each unit fraction together to make a whole as you discuss the addition problem. This type of problem (unit addition problems) should be explored with various fractions with common denominators. The next step is to explore adding fractions with the same denominators, which are larger than a unit fraction. For example, 1/4 + 2/4 = 3/4. A model is located in Figure 9–1.

Use circle fraction pieces to explore these problems as well. Ask students if the denominator changed. When they note these did not change, ask them why. Guide students to see that the denominator is how many unit fractions it takes to make a whole and that each problem they have explored has the same number of unit fractions to make a whole. It is also important at this time to use other fraction area models such as pattern blocks to see that the equation produces the same fractional answer, regardless of the manipulative used. For example, while students can explore 1/3 + 2/3 = 1 whole with circle fractions, they can also use the hexagon as a whole and find that 1/3 + 2/3 = 1, with the rhombus serving as 1/3 of the whole. Students can also use the trapezoid as a whole and explore how 1/3 + 2/3 = 1 whole trapezoid, with the triangles serving as thirds. Using fraction circles, have students show 1/5 and 2/5. Ask the students how they could add these together. This will show 3/5 of the circle is now complete. Ask students how many more pieces would be needed to make a whole? (The circle fraction pieces are useful because the whole is easy to determine.) If students seem to understand this concept of making a whole, follow up by asking what they might do if they have a whole but then need to add 2/5 more. See if students can identify that this would be 1 2/5. If students are still struggling to make a whole, try with different fractions. Remember to have the whole circle fraction piece set (1/1) on the table for students to refer to when you discuss the whole. As students contemplate adding two fractions with like denominators, ask them what they would do to add two whole numbers together. This questioning helps students see that they are doing the same thing when adding fractions, combining or putting together. As students explore concepts of addition of fractions with like denominators using circle fractions, it is important to also explore subtraction and to give contextual problems that students can use the circle fractions to solve. For example, tell students you have 2/8 of the pepperoni pizza left and 3/8 of the cheese pizza and then ask students to use the fraction circles to determine how much of a whole pizza would there be if these were put together. Similar problems in which students give pieces of pizza away help explore subtraction. However, you don't only want to explore pizza. Try to give a variety of problems that might involve

Present Problem

$$\frac{1}{4} + \frac{2}{4} =$$

Make one fourth using rectangle blocks.

Make two fourths using rectangle blocks.

Combine one fourth and two fourths using one rectangle and complete problem.

$$\frac{1}{4} + \frac{2}{4} = \frac{3}{4}$$

Figure 9–1. Model of Addition Using Fractions Larger Than a Unit.

circle pieces. As mentioned in the previous chapter, pattern blocks are also useful area models, as are base 10 blocks. Each of these models helps students explore and generalize understandings about adding and subtracting fractions with like denominators. It is also important to explore adding like denominators with the length model before moving to addition with different denominators. See details about this model in the length model section.

Before moving to adding fractions with different denominators, teachers should guide students to explore subtraction of fractions following the inverse of the model above. For example, there was 3/4 of a pizza and I ate 2/4 of it. How much pizza is left? Allow students to make the connection of how and why the denominator doesn't change, because the parts that are needed to make a whole remain the same throughout the problem. As contextual problems may be given, it is important that the contextual problems match the manipulatives. For example, the circle fractions would be appropriate when exploring pizzas and pies. Throughout this discussion, help students find how addition and subtraction of fractions relates to addition and subtraction of whole numbers. Share contextual problems with whole numbers and then share the same problems with fractions so students see how these relate.

The next phase of addition and subtraction of fractions involves adding fractions without common denominators. It is important that students can estimate and make sense of problems that involve denominators that are not the same. Begin with manipulatives to solve problems such as: *At the party, Julia was given a small cake to decorate. She decorated and iced 1/3 of it and allowed Juan to ice and decorate 1/6 of her cake. Julia's sisters prefer cake without icing, so she left the rest of the cake plain. How much of the cake was decorated?* Write the problem on the board (1/3 + 1/6) after students have shared what type of computation is needed (addition). While students can see with the manipulatives (placing a 1/3 piece and 1/6 piece next to each other) that 1/2 of the circle is full; they will also see that the pieces are all different sizes. Have students share the answer, and many will share that the answer is 1/2. Ask students how they decided 1/2. While it is important to recognize students for reasoning that 1/2 is used and that they were able to formulate and solve the problem with manipulatives, it will be important to note that they will not always be able to visually see the answer. Also, it is important to emphasize that different denominators are being used and question to ensure students remember that the thirds would take three pieces to be one whole, sixths would take six pieces to make one whole, and the answer 1/2 means that two of those pieces would be needed to make one whole.

Next, teachers will give a problem in which the answer cannot be easily determined by using manipulatives and visually examining the problem such as 1/4 + 2/6 = 7/12. While students will not know the exact answer simply by using manipulatives, help them determine that it is just slightly over 1/2. Focusing on estimation is an important element in sense-making for students. Then, with fractional pieces for twelfths, the teacher could ask students to find a fractional set that can use unit fractions to exactly cover the same amount as the 1/4 + 2/6. They will be able to see how 7/12 makes the same amount as this. Students need multiple experiences with estimating and exploring these addition problems. While they may not have the complete tools for solving how to find the common denominator, these experiences are strengthening their confidence in sense-making in fractions, strengthening the idea that addition and subtraction of fractions follows the same basic premise as addition and subtraction of whole numbers. It also allows students to see they need further understanding to be able to add all fractions with different denominators.

Length

Addition with the length model involves using fraction bars to add various fractional amounts. Using fraction bars with attention to unit fractions is helpful in exploring addition. These fraction bars can be used in conjunction with a number line so students are able to transition easily to the representational/semi-concrete model with just the number line. This exploration can be concrete as students line up fraction bars and units next to the number line in order to make sense of the process of adding on a number line. Students need opportunities to explore, with teacher guidance, adding like denominators with the length model before moving to addition with different denominators. Having the whole is important when using fraction bars. Subtraction should be introduced after addition is understood at each phase. An example of modeling addition using a number line at the concrete level is shown in Figure 9–2.

Concrete examples should begin with fractions less than one, but fractions with value greater than one need to be intertwined to avoid the misconceptions that all fractions are less than one. It is important to also provide real-world scenarios and context where area fractions might occur. For example, a portion of a garden, pizza, classroom, or measurement unit could be used as examples.

ADDITION AND SUBTRACTION OF FRACTIONS AT THE REPRESENTATIONAL/SEMI-CONCRETE LEVEL

Representations will be very similar to the concrete models. However, it is important to begin with prepartitioned images. This allows students to begin by thinking about the addition aspect of the fraction without the added demands associated with drawing. After students are comfortable with the prepartitioned elements, it is important to push students to then create their own representational/semi-concrete representations of fractions. Students have a difficult time with circular partitioning, so having them draw rectangular shapes is one way to help students explore this concept. See Figure 9–3 for an example intervention instructional segment of a representational/semi-concrete level of addition with the area model of fractions within real-world contexts.

$$\frac{1}{3} + \frac{2}{3} =$$

Present the problem. Place one third on the number line. *I start by making one third. When the denominator is 3, there are three parts in the whole, so I look for blocks that will make a whole with three. How many of these blocks will it take to make a whole? Yes, three. What is the numerator? Yes, the numerator is 1, so, I take one and place it on the number line.*

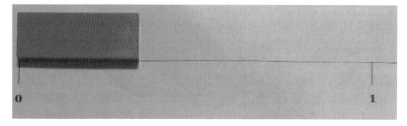

Discuss addition. *This sign means that I will add, so I will combine one third with two thirds. I make two thirds with two blocks.*

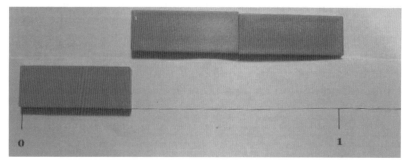

Place two thirds on the number line. *I am adding, so I combine one third and two thirds on the number line. Now we have three thirds.*

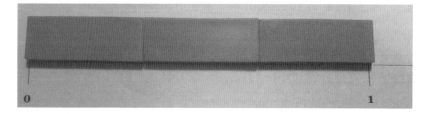

Write the answer. *Count with me: how many thirds do I have? 1 . . . 2 . . . 3 . I notice that my whole has three parts and I used three of them. I write three thirds after the equal sign. My denominator is 3 and the numerator is 3.*

$$\frac{1}{3} + \frac{2}{3} = \frac{3}{3}$$

Figure 9–2. Modeling and Guided Practice Addition Using a Number Line at Concrete-Level Modeling. continues

Guided Practice

$$\frac{2}{5} + \frac{2}{5} =$$

Present the problem. Place two fifths on the number line. *We start by making two fifths. When the denominator is 5, how many parts are in the whole? Yes, five, so we look for blocks that will make a whole with how many? Yes, five. Now we have fifths. What is the numerator? Yes, two. So how many should we put on the number line? Yes, two.*

Discuss addition. *What does the sign tell us to do? Yes, add, so we will combine two fifths with two fifths. Make two fifths with two blocks.*

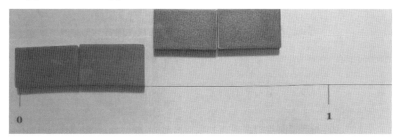

Place two fifths on the number line. *We are adding, so we combine two fifths with two fifths on the number line.*

Write the answer. *Count with me how many fifths do we have? 1 . . . 2 . . . 3 . . . 4. Notice that our whole still has five parts. How many did we use? Yes, four. Write four fifths after the equal sign. What is our denominator? Yes, five. What is our numerator? Yes, four.*

$$\frac{2}{5} + \frac{2}{5} = \frac{4}{5}$$

Figure 9–2. continued

Problem	Think Aloud and Represent the Problem
Annie ate 2/8 of the pizza and Joseph ate 1/2 of the pizza. How much of the pizza was eaten? (area)	*What does this problem ask? (The problem asks how much of the pizza was eaten by these two people.)* *What do we know about a pizza? How can we show how much Annie ate?* *How can we show how much Joseph ate? Can you draw a picture to show me this?* *What could we do to put these together and determine how much of the pizza was eaten by both people? Now that you have drawn it, how can you express it in terms of fractions? How do you determine how much of the whole we have marked?*
Julia's ribbon was 1/2 foot long. Jerome's ribbon was 2/8 foot long. How long would their two ribbons be if they were lined up together? (length)	*What does this problem ask? (The problem asks how long the two pencils would be if they were lined up together.)* *How can you show how long Julia's ribbon is? How can you show how long Jerome's ribbon is? How could we put them together? So we have a drawing of them together, but how can we express this in terms of a fraction? How much of the whole is marked? Do you think this is more or less than one foot? Why?*

Figure 9–3. *Representation-Level Instruction of Addition With the Area Model of Fractions in Context.*

Area

One difficult point for students in addition of fractions is determining how to add fractions with unlike denominators. This is especially difficult for students who are still struggling with basic multiplication facts and factors. One way to help students see how and why common denominators are important is through the area model. Selecting fractions with smaller denominators is important to bring the focus to the process, rather than the counting and drawing of lines. For example, one problem to begin with would be 1/3 + 1/4. Teachers should begin by having students draw two squares. The first square should have 1/3 drawn with horizontal lines and the second square should have 1/4 drawn using vertical lines. While conceptually, it doesn't matter if you have vertical or horizontal lines, consistency in representations can be useful to avoid confusion. After both figures have been drawn, ask students what it means to add and how they might add these together. Students should share that adding is combining these two amounts. However, most predictions for adding these two will involve estimates. As a teacher, it is vital that students realize we need a precise answer. Point out to students that the two fractions do not have equal-sized pieces and we need to find a way to have equal pieces so we can combine them, just as we needed equal-sized pieces when comparing fractions (see previous chapter). Ask students if they have an idea of how we could create equal-sized pieces. If they are unsure, you can show them that by adding the horizontal lines to the second shape with vertical lines (to make 4/12) and vice versa (to make 3/12) (Figure 9–4). Ask students if the shaded area has changed in each shape (no). It is important that the teacher helps students see the same amount is shaded, but we are expressing it in different terms so the two amounts can be added. It is also critical to ensure that students are still aware that the rectangle is the whole (which now has a total of 12 pieces). Ask the students if they were to put the 4/12 and 3/12 together, how much would there be? (7/12). After this example, students need additional teacher-guided and independent practice. In addition to problems such as the one above, students need to be pushed to explore problems where the answer is greater than one and subtraction problems. Multiple examples of these need to be practiced. These activities will form the foundation for abstract equations. For example, 1/3 + 1/4 = 1/3 (4/4) + 1/4 (3/3) = 4/12 + 3/12 = 7/12 shows how factors play a role in creating common denominators (see Figure 9–4).

Length Model

The length model should be introduced after students are familiar with the area model. When adding like denominators, students can easily equipartition a number line and find the total number. It is important that the number line benchmarks that may be given are varied. In other words, students should not always see a number line with the markings for 0 and 1 only or they won't know what to do when they see a number line that shows a 0, 1, and 2. Provide multiple benchmark markings and explicitly discuss them.

Addition of fractions with unlike denominators focuses more on the role factors play in finding common denominators. Provide multiple number lines so students can explore the meaning of both addends and discuss ways to equipartition so they can build upon each other. Ask questions that help students realize the denominators are key in determining the next steps.

DEVELOPING UNDERSTANDING LEADING TO PROCEDURAL KNOWLEDGE RELATED TO UNLIKE DENOMINATORS

Prior to the abstract level of operations, especially those that involve unlike denominators, students must have conceptual knowledge that is developed at the concrete and representational/semi-concrete. To this point, the emphasis of instruction has been exploration, estimation, and development of sense-making, all skills that are essential to understanding how to manipulate fraction numbers. Provide students with experiences in which they can

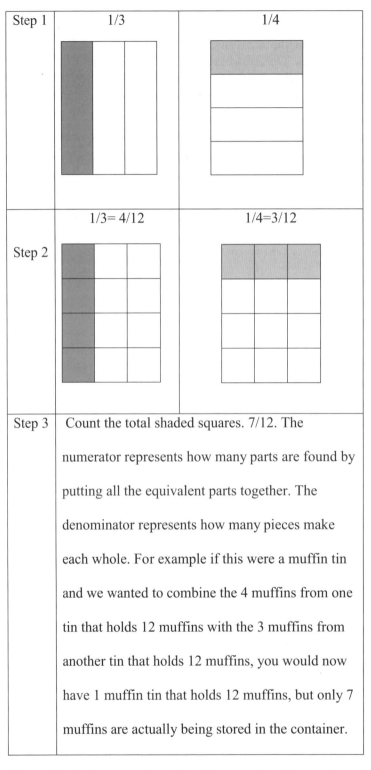

Figure 9–4. Representational/Semi-Concrete Level Instruction of Addition With Area Model of Fractions.

physically see how fractions can be transformed into other fractions that have the same value using multiplication, specifically multiplication by one. Just like the underpinnings of addition of fractions includes the combination of unit fractions (2/3 = 1/3 + 1/3), finding equivalent fractions begins by showing how to make fractions using multiplication sentences. Prior to learning about procedures related to making equivalent fractions, students must understand that fractions can be made using multiplication (3/6 = 1/3 × 2/2). At the concrete level, the lesson begins with the presentation of a partial equation, three sixths equals an unknown number of halves (3/6 = ?/2). Then, the teacher and students use fraction blocks to find the unknown. Once the two sets of blocks representing 3/6 and 1/2 are shown together, begin a discussion about how many sixths comprise one half. This should lead to the realization that it takes three sixths for each one half. Then this information is used to make a multiplication sentence showing the transformation (3/6 = 1/2 × 3/3). The same process should be followed for fractions larger than one (e.g., 12/10 = ?/5). Fraction blocks as well as tiles and number line should be used so that instruction continues consistently with area and length models. These processes are demonstrated using explicit instruction in Figure 9–5.

At the representational/semi-concrete level, present students with shapes with predrawn equipartitions and instead of putting blocks together, shade portions of the two shapes. Have the same discussion about how many parts it takes to make each unit of the other fraction and use this information to make a multiplication sentence. Using number lines, provide students with two marked number lines and have students shade each portion of the corresponding number line. These processes are demonstrated using explicit instruction in Figure 9–6.

CRA/CSA APPLICATION FOR ADDITION AND SUBTRACTION OF FRACTIONS AT THE ABSTRACT LEVEL

Now that students have had opportunities to explore addition of fractions using various representations, they should be ready to explore these concepts in the abstract phase. Revisiting problems such as the one described above are critical in understanding how to represent and solve this abstractly. For example, writing 1/3 + 1/4 = 1/3 (4/4) + 1/4 (3/3) = 4/12 + 3/12 = 7/12 and discussing how this was modeled in the representational/semi-concrete phase and why you can multiply a fraction by 4/4 or 3/3 (the identity property) helps students understand the connection to whole-number operations they have learned in earlier grades.

Model Equivalent Fractions Using Area Model

Present problem. *I have three sixths and it is equal to a fraction with a denominator of 2.*

Make completed fraction. *First, I am going to make three sixths. I need to find the blocks that have six parts in the whole. The numerator tells how many of these sixths I need. What is the numerator? Yes, 3. So I put three sixths together.*

Make equivalent fraction. *I am going to find the fraction that is equal. The denominator is 2, so I need to find the blocks that have two parts in the whole. The numerator tells how many I need. So, I find the one half blocks and see how many it takes to equal three sixths. It takes one, so the numerator is 1. I compare the two to make sure that they are the same.*

Figure 9–5. Making Equivalent Fractions at Concrete Level. continues

Now I can write the complete equivalent fraction, one half.

$$\frac{3}{6} = \frac{1}{2}$$

Write Equation. *I can write a multiplication equation to show that these two fractions are equal. I look at how many sixths it takes to make one half. Count with me: 1 . . . 2 . . . 3. For each one half, I need three sixths. It takes three times the numerator and three times the denominator to turn one half into three sixths. If I write that equation, it is three sixths equals one half times three thirds. The fraction three thirds equals 1. What happens to any number when it is multiplied by 1, like 4 times 1 or 6 times 1? The number stays the same. Is one half the same as three sixths? Yes.*

$$\frac{3}{6} = \frac{1}{2} \times \frac{3}{3}$$

Figure 9–5. continues

Guided Practice in Making Equivalent Fractions Using the Area Model

Present problem. *I have two thirds and it is equal to a fraction with a denominator of 6.*

$$\frac{2}{3} = \frac{}{6}$$

Make completed fraction. *First, we make two thirds. We need to find the blocks that have how many parts in the whole? Yes, three. The numerator tells how many of these thirds we need. How many thirds should we put together?*

Make equivalent fraction. *Now we find the fraction that is equal. The denominator is 6, so we need to find the blocks that will combine to make a whole with how many? Yes, six. We do not know that numerator, so we put sixths together until our fractions are equal. How many sixths does it take? Yes, four. So the numerator is what? Yes, 4. Compare the two fractions to make sure that they are the same.*

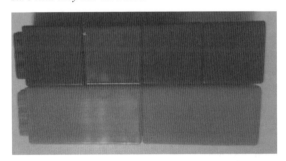

Now I can write the complete equivalent fraction, four sixths.

$$\frac{2}{3} = \frac{4}{6}$$

Figure 9–5. continues

Write equation. *We can write a multiplication equation to show that these two fractions are equal. Look at how many sixths it takes to make each third. Count with me: 1 . . . 2. For each one third, how many sixths do we need? Yes, two. It takes two times the numerator and two times the denominator to turn two thirds into four sixths. If we write that equation, it is four sixths equals two thirds times two halves. Is two halves the same as 1? Yes. What happens to any number when it is multiplied by 1, like 3 times 1 or 5 times 1? The number stays the same. Is two thirds the same as four sixths? Yes.*

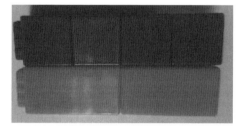

4	=	2	×	2
6		3		2

Figure 9–5. continues

Guided Practice in Making Equivalent Fractions Using the Length Model

Present problem. *I have two thirds and it is equal to a fraction with a denominator of 6.*

$$\frac{2}{3} = \frac{}{6}$$

Make completed fraction. *First, we make two thirds. We need to find the blocks that have how many parts in the whole? Yes, three. The numerator tells how many of these thirds we need. How many thirds should we put together?*

Make equivalent fraction. *Now we find the fraction that is equal. The denominator is 6, so we need to find the blocks that will combine to make a whole with how many? Yes, six. We do not know that numerator, so we put sixths together until our fractions are equal. How many sixths does it take? Yes, four. So the numerator is what? Yes, 4. Compare the two fractions to make sure that they are the same.*

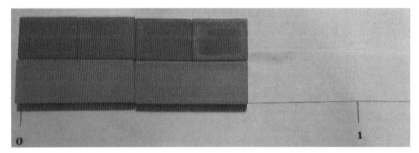

Now I can write the complete equivalent fraction, four sixths.

$$\frac{2}{3} = \frac{4}{6}$$

Figure 9–5. continues

Write equation. *We can write a multiplication equation to show that these two fractions are equal. Look at how many sixths it takes to make each third. Count with me: 1 . . . 2. For each one third, how many sixths do we need? Yes, two. It takes two times the numerator and two times the denominator to turn two thirds into four sixths. If we write that equation, it is four sixths equals two thirds times two halves. Is two halves the same as 1? Yes. What happens to any number when it is multiplied by 1, like 3 times 1 or 5 times 1? The number stays the same. Is two thirds the same as four sixths? Yes.*

Figure 9–5. continued

Model Equivalent Fractions using Area Model

Present problem. *I have four eighths and it is equal to a fraction with a denominator of two.*

$$\frac{4}{8} = \frac{}{2}$$

Make completed fraction. *First, I am going to make four eighths. I need to find the shape that has eight parts in the whole. The numerator tells how many of these eights I use. What is the numerator? Yes, four. So I shade four parts.*

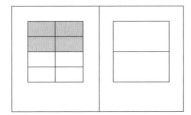

Make equivalent fraction. *I am going to find the fraction that is equal. The denominator is two, so I need to find the shape that has two parts in the whole. The numerator tells how many I need. So, I look to see how many halves are the same as four eighths. It takes one] shade that part. The numerator is one. I compare the two fractions to make sure that they are the same.*

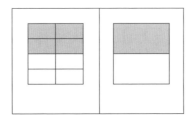

Now I can write the complete equivalent fraction, one half.

$$\frac{4}{8} = \frac{1}{2}$$

Figure 9–6. *Making Equivalent Fractions at Representational/Semi-Concrete Level.* continues

Write Equation. *I can write a multiplication equation to show that these two fractions are equal. I look at how many eighths it takes to make one half. Count with me...1..2..3...4. For each one half, I need four eighths. It takes four times the numerator and four times the denominator to turn one half into four eighths. If I write that equation, it is four eighths equals one half times four fourths. The fraction four fourths equals one. What happens to any number when it is multiplied by one, like six times one or two times one? The number stays the same. Is one half the same as four eighths? Yes.*

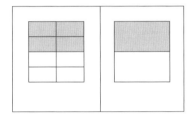

$$\frac{4}{8} = \frac{1}{2} \times \frac{4}{4}$$

Figure 9–6. continues

Guided Practice in Making Equivalent Fractions using the Area Model

Present problem. *I have two thirds and it is equal to a fraction with a denominator of six.*

$$\frac{2}{3} = \frac{}{6}$$

Make completed fraction. *First, we make two thirds. We need to find the shape that has how many parts in the whole? Yes, three. The numerator tells how many of these thirds we need. How many thirds should we shade?*

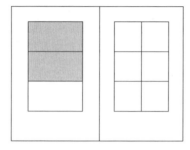

Make equivalent fraction. *Now we find the fraction that is equal. The denominator is six, so we need to find the shape that has how many parts in the whole? Yes, six. We do not know that numerator, so we shade sixths until our fractions are equal. How many sixths does it take? Yes, four. So the numerator is what? Yes, four. Compare the two fractions to make sure that they are the same.*

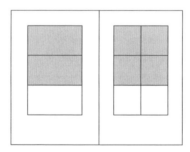

Now I can write the complete equivalent fraction, four sixths.

$$\frac{2}{3} = \frac{4}{6}$$

Figure 9–6. continues

Write equation. *We can write a multiplication equation to show that these two fractions are equal. Look at how many sixths it takes to make each third. Count with me...1..2. For each one third, how many sixths do we need? Yes, two. It takes two times the numerator and two times the denominator to turn two thirds into four sixths. If we write that equation, it is four sixths equals two thirds times two halves. Is two halves the same as one? Yes. What happens to any number when it is multiplied by one, like three times one or five times one? The number stays the same. Is two thirds the same as four sixths? Yes.*

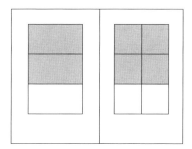

Figure 9–6. continues

Guided Practice in Making Equivalent Fractions using the Length Model

Present problem. *I have two thirds and it is equal to a fraction with a denominator of six.*

$$\frac{2}{3} = \frac{}{6}$$

Make completed fraction. *First, we make two thirds. We find the number line that shows thirds; how many parts should be between zero and one? Yes, three. The numerator tells how many of these thirds we will shade. How many thirds should we shade?*

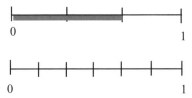

Make equivalent fraction. *Now we find the fraction that is equal. The denominator is six, so we need to find number line that has how many parts between zero and one?? Yes, six. We do not know that numerator, so we shade sixths until our fractions are equal. How many sixths does it take? Yes, four. So the numerator is what? Yes, four. Compare the two fractions to make sure that they are the same.*

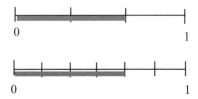

Now I can write the complete equivalent fraction, four sixths.

$$\frac{2}{3} = \frac{4}{6}$$

Figure 9–6. continues

Write equation. *We can write a multiplication equation to show that these two fractions are equal. Look at how many sixths it takes to make each third. Count with me...1..2. For each one third, how many sixths do we need? Yes, two. It takes two times the numerator and two times the denominator to turn two thirds into four sixths. If we write that equation, it is four sixths equals two thirds times two halves. Is two halves the same as one? Yes. What happens to any number when it is multiplied by one, like three times one or five times one? The number stays the same. Is two thirds the same as four sixths? Yes.*

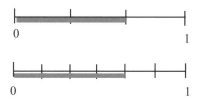

Figure 9–6. continued

CRA/CSA APPLICATION FOR MULTIPLICATION OF FRACTIONS

Multiplication of fractions is often taught procedurally by teachers, because it is a simple procedure (multiply across). However, when students do not conceptually understand why this works or what they are doing when multiplying fractions, they can easily confuse this algorithm with others. Often, after students have been taught to multiply across, they will follow this procedure for addition, because they perceive addition as the easiest algorithm. CRA/CSA helps build conceptual knowledge and avoid errors due to this type of novice conception.

CRA/CSA APPLICATION FOR MULTIPLICATION OF FRACTIONS—CONCRETE

When beginning multiplication with an intervention group, it is important to review whole-number multiplication. A concrete way to review this concept before extending it to fractions is to ask students to create equal groups with whole items and find the total number of items. It is important that students are able to see that multiplication of fractions is finding the same thing as multiplication of whole numbers. This is often confusing for students because they are taught that multiplication problems have larger answers, and with fractions, the solutions are actually smaller. However, this is because you are making the fraction into equal fractional parts, which would create a smaller solution, but still it is multiplication. A model of multiplication of fractions is shown in Figure 9–7.

The physical movement of the whole-number multiplication of fractions is important as you transition students to the representation stage of

Problem: $\frac{1}{2} \times \frac{1}{3}$
Translate problem: There are one-half groups of one third.
Represent $\frac{1}{3}$ There are three parts in the whole and one part is used.
Represent the problem by making equal groups that are one half the size. Now, there are six parts in the whole.
Represent the answer. $\frac{1}{6}$ There are six parts in the whole and one part is used.

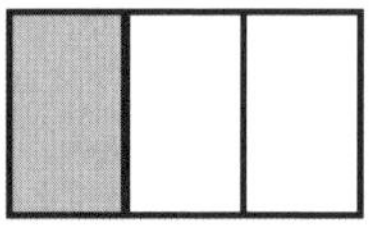

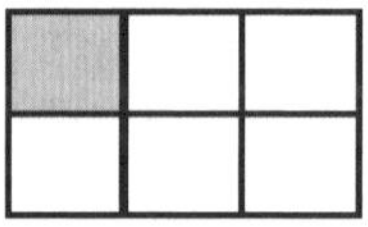

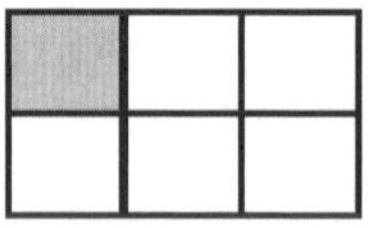

Figure 9–7. Model for Multiplication of Fractions.

multiplication on fractions. In addition, students should explore contextual problems with manipulatives problems such as:

1. I have 3 pans with 4/5 of a whole pizza on each pan. How many pizzas do I have?
2. You have 3/4 of a case of 12 coke cans. How many coke cans do you have?

Area Model

Providing students with manipulatives to interpret and solve these problems is critical in developing conceptual understating of the process. While you can explore multiplication of fractions with concrete models, there are only a limited number that can be explored in a way where the benefits of the concrete outweigh the drawbacks of the steps and chances for error. Teachers could use 1-inch tiles on hundreds chart board or a geoboard to solve a problem such as 1/4 × 1/2. Students begin by determining the whole (such as a whole geoboard or creating a square on the geoboard). As teachers guide students in creating the whole, it is important to ensure the geoboard or tile wholes are done in a way where they can be divided into the parts necessary for the multiplication to be successful. So for example, you would not want to choose a 3 × 3 square for the problem 1/4 × 1/2. For the example 1/4 × 1/2, teachers can encourage students to make a 2 × 4 array with either tiles or the geoboard. This array is their "whole" shape. Now they are going to show 1/2 of this shape. So they could take the 2 × 4 on the geoboard and use a rubber band to make a 1 × 4 piece, which is half of the whole shape. Now that they have 1/2, the problem asks them to find 1/4 of that 1/2 (as the problem asks). So they will look at the 1 × 4 array and determine what 1 of 4 groups would be. This can be marked with a different rubber band. Now ask students to share what part of the whole 1/4 of the 1/2 was (1/8). Ask them to share where the 8 came from in this problem and scaffold as needed (Figure 9–8).

Length Model

Similar to the area model, choosing a problem is essential in effectively using the concrete model to explore the conceptual meaning of multiplication of fractions. For example, begin with 1/3 × 1/2. Have students use their fraction bars to line up a bar that shows 1/2 out of 1 on the number line. Explain to students that this is what we begin with in the problem (1/2). Now they are challenged to find 1/3 of that 1/2. So ask students to use their fraction bars and find which three bars are the exact same length as the 1/2 bar (three

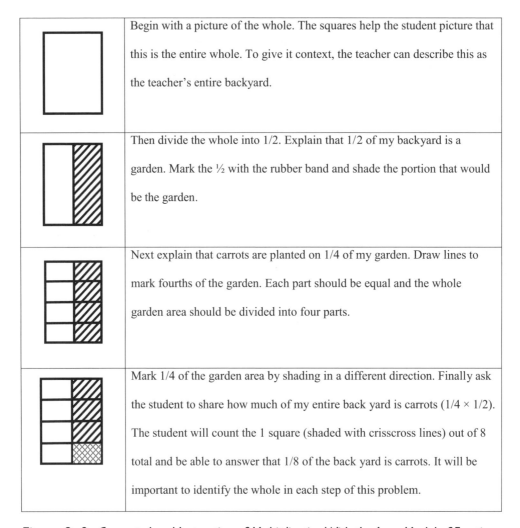

Figure 9–8. Concrete-Level Instruction of Multiplication With the Area Model of Fractions.

of the sixth bars). The problem asks what is 1/3 of the 1/2, so students will share that the 1/6 covers 1/3 of the 1/2 bar. While this is confusing in language, the manipulatives help students see what is actually occurring, groups that are one third the size of one half. Instructors should then ask students to predict what 2/3 of 1/2 would be (two sixths). Instructors ask students to explain the difference between the two problems. It is important that a record of problems solved by students is visible on something such as chart paper, so that students are able to look for a pattern of these discoveries. Explore this process with a few problems so that eventually students can answer without teacher scaffolding. This indicates that they are ready to move to the representational/semi-concrete phase.

CRA APPLICATION FOR MULTIPLICATION OF FRACTIONS—REPRESENTATIONAL/SEMI-CONCRETE

Area

The representational/semi-concrete phase of multiplication of fractions allows students to see how illustrations can help solve problems involving multiplication of fractions. Posing a problem such as 2/5 × 2/3 is a bit easier with drawings versus concrete manipulatives. To transition to the drawings, teachers can use transparencies that represent each of the fractions. Begin with the transparency that shows 2/3 shaded horizontally. Then have students find the transparency that shows 2/5 of the whole vertically. Now ask students to remember what they did with the concrete models. Are they really looking for 2/5 of the whole or are they looking for 2/5 of 2/3 (the latter)? So when the transparency 2/5 is placed over 2/3, they are looking at the 2/5 of the 2/3 (where both colors are shown). Therefore, they should see that four pieces are colored, but these pieces are 4/15 of the whole. Do a few more with transparencies and then transition to drawing the problems. Ask students to predict why it is important to equipartition the whole with the second fraction when you are only using it within the first fractions. It is important that they realize the final answer shares how many pieces of the whole are found from this problem, so the whole needs to be equipartitioned. See Figure 9–9 for an illustration of the representational/semi-concrete phase of multiplication with the area model. Teachers should question students to ensure they understand why the denominator changes and why the answer is where the two fractions overlap. It is also important before moving to another model that students have opportunities to explore fractional concepts of finding common denominators, adding, subtracting, and multiplying in a practice setting where all problem types are given. This allows the teacher to see if there are elements of confusion between the problem types. If students lack conceptual understanding of why they are doing each step, they can easily confuse the models for each of these, just as they are confuse the traditional algorithms that have been taught in the past.

Length

Once students demonstrate understanding of the representational/semi-concrete-level instruction of multiplication with the area model of fractions, instructors should give students another model to provide more complete understanding of this process. Follow similar procedures to the concrete level of instruction for the length model of multiplication of fractions. However, students will not have access to the fraction bars. Therefore, initially, number lines need to be equipartitioned in ways that lead students to finding common

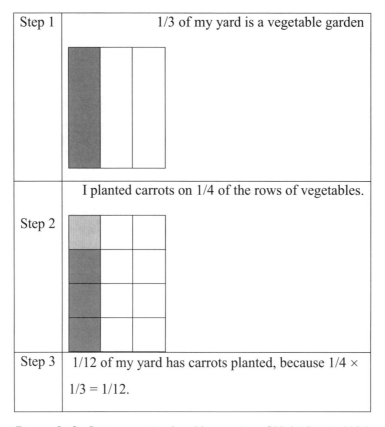

Figure 9–9. Representation-Level Instruction of Multiplication With the Area Model of Fractions (1/4 of 1/3 or 1/4 × 1/3).

denominators. When students are transitioned to number lines that are not equipartitioned, teachers need to provide problems that are easier to draw and solve on a number line.

CRA APPLICATION FOR MULTIPLICATION OF FRACTIONS—ABSTRACT

When moving to the abstract phase, the teacher can display the chart of equations that has been built from the previous phases (Figure 9–10). Ask students to find a pattern as they search these problems. Then ask them to try to use the prior problems to predict if 1/5 × 2/6 will be larger or smaller than 1/5. Ask students to predict the way to solve this. With a large list of equations displayed, the teacher should be able to guide students in discovering that the numerators and the denominators can be multiplied to give the solution. Then check this with the area model so students see the abstract pattern still can be verified with the models previously learned. Try a few more equations and check them with the representational/semi-concrete

1/2 × 1/3 = 1/6	1/4 × 4/3 = 4/12
1/4 × 1/3 = 1/12	5/4 × 4/3 = 20/12
3/4 × 1/3 = 3/12	3/4 × 4/3 = 12/12
5/4 × 1/3 = 5/12	5/4 × 5/4 = 25/16

Figure 9–10. Chart of Multiplication Equations.

phase of the area model. Then have students try problems on their own. Just like the representational/semi-concrete phase, it is important to end by practicing addition and multiplication problems together. In the small group, encourage students to share what they are doing at each phase of these problems.

DIVISION OF FRACTIONS

At the elementary level, students do not learn to divide a fraction by a fraction. However, they do explore dividing a unit fraction by a whole number and a whole number by a unit fraction. Similar to other operations, it is important for teachers to begin with what students already know. Teachers review how fractions are division problems and then review what division of whole numbers means. For example, they could ask students to solve the problem, "How many 2-inch strips of construction paper can I make if I am cutting along the side of a 12-inch-long piece of construction paper?" Students should solve this problem by actually using construction paper and rulers to solve it. After reviewing division of whole numbers, teachers should give similar contextual problems, but with one of the numbers being a fraction. For example, "How many 1/2-inch strips of construction paper can I make if I am cutting along the side of a 12-inch-long piece of construction paper?" Ask students if they predict the answer will be larger or smaller and why. Then the teacher should support students as they test their predictions.

10 ÷ 2 means how many groups of two are in the number 10. 1 ÷ 1/4 means how many one fourths are in one. Students need to see how dividing with fractions is similar to division of whole numbers, but they also need to understand why the answer will be greater when dividing with a unit fraction. Then students explore how many one fourths are in the number 2. Then how many one fourths are in the number 4? Finally, students explore how many one fourths are in the number 10. The teacher can see if students are able to utilize their understanding of the first problem to solve these follow-up questions. As they solve these problems, students should be illustrating the problem, similar to Figure 9–11.

2 ÷ 1/4	There are 4/4 in 1, so there are 8/4 in 2.
4 ÷ 1/4	There are 4/4 in 1, so there are 16/4 in 4.
10 ÷ 1/4	There are 4/4 in 1, so there are 40/4 in 10.

Figure 9–11. *Division of a Whole Number by a Unit Fraction.*

Fifth graders are also challenged to solve division of a unit fraction by a whole number. Drawing contextual pictures is helpful as students make sense of these problems as well. For example, I have 1/2 of my sandwich left and am splitting it with four people. How much of a sandwich will each person receive? Have students draw the whole sandwich and shade the 1/2 of the sandwich that is gone. Then they will divide the two halves of the sandwich (shaded and unshaded) into four pieces. As a teacher, it is important to remind students that they are now answering how much of a sandwich each person receives (1/8). Intervention needs to be focused on the prior content and connecting division to their understanding of fractions and whole-number division. These concepts are more important than the actual process of solving these equations, because while division groundwork is explored in fifth grade, it will be revisited in greater detail in sixth grade. To this point, students will be expected to fully understand and complete operations with addition, subtraction, and multiplication of fractions; students explore division and make models rather than use procedures and algorithms in fifth grade.

USING UNDERSTANDING OF FRACTIONS TO DEVELOP UNDERSTANDING OF BASIC DECIMAL OPERATIONS

Another concept introduced in late elementary grades is decimals, another way of representing fractions. When completing operations such as addition and subtraction of decimals, it is important to teach conceptual understanding prior to procedural knowledge. Procedurally, students align decimals and complete the operation; however, students who struggle with mathematics need explicit instruction as to how and why this procedure is commonly accepted.

Decimals are fractions with denominators of 10, 100, 1,000, and so on, and they are written using decimal point and number to the right of zero. Initial instruction in decimals develops students' understanding that fractions with denominators of 10 and 100 can be written another way (5/10 = 0.5 and 6/100 = 0.06). With this knowledge, students can begin by making equivalent fractions or decimals using the same process described earlier in this chapter. See Figure 9–12 for an example of explicit modeling of this concept.

Present problem. *I have three tenths and it is equal to a fraction with a denominator of one hundred. If I wrote the fraction as a decimal, three tenths is written as zero, point three.*

$$\frac{3}{10} = \frac{}{100}$$

$$0.3 \quad = \quad ?$$

Make completed fraction. *First, I am going to make three tenths. I need to find the shape that has ten parts in the whole. The numerator tells how many of these tenths I need. What is the numerator? Yes, three. So I fill three of the tenths with blocks.*

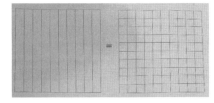

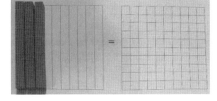

Make equivalent fraction. *I am going to find the fraction that is equal. The denominator is one hundred, so I need to find the shape that has one hundred parts in the whole. The numerator tells how many I need. So, I use the hundredths blocks to fill in the same amount at three tenths. How many hundredths will it take to fill one column? Yes, ten.*

Figure 9–12. *Equivalent Fractions Leading to Equivalent Decimals.* continues

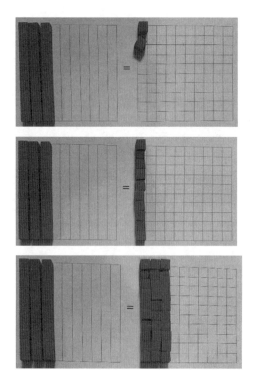

Now I can write the complete equivalent fraction, thirty hundredths. I can write this fraction as a decimal. Three tenths is zero, point three. Thirty hundredths is zero, point, three, zero.

$$\frac{3}{10} = \frac{30}{100}$$

$$0.3 = 0.30$$

Write Equation. *I can write a multiplication equation to show that these two fractions are equal. I look at how many hundredths it takes to make one tenth. Count with me…1..2..3…4…5…6…7…8…9…10 For each one tenth, I need ten hundredths. It takes ten times the numerator and ten times the denominator to turn three tenths into 30 hundredths.*

Figure 9–12. continues

If I write that equation, it is three tenths times ten tenths equals thirty hundredths. The fraction ten tenths equals one. What happens to any number when it is multiplied by one, like four times one or six times one? The number stays the same. Is three tenths equal to thirty hundredths? Yes.

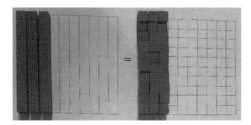

$$\frac{30}{100} = \frac{3}{10} \times \frac{10}{10}$$

Figure 9–12. continued

After students can make equivalent fractions, teach them how to add fractions with unlike denominators such as 10 and 100 using concrete and representational models. These models will demonstrate why 0.4 and 0.05 do not equal 0.9 and 0.09 because 4/10 and 5/100 is 45/100 (40/100 + 5/100). See Figure 9–13 for an example of instructional steps for teaching this process at the concrete and representational levels.

CHAPTER SUMMARY

Computation of fractions can be disconnected from whole-number computation when taught without conceptual understanding and models. It is easy for students to confuse processes and meanings because they are applying misconceptions about computation gained when exploring whole-number computation in earlier grades. Therefore, it is important to go slowly, to explore multiple models, and to ensure students are able to explain the connection between computational procedure with fractions to the same procedure with whole numbers. Provide an instructional trajectory that systematically builds students' understanding of operations, first with unit fractions and then fractions with larger numerators and like denominators.

Begin with problem and translate fraction numbers into decimal notation.

$$\frac{3}{10} + \frac{4}{100} = \frac{}{100}$$

0.3 + 0.04 =

Use tables like the ones below to organize concrete materials.

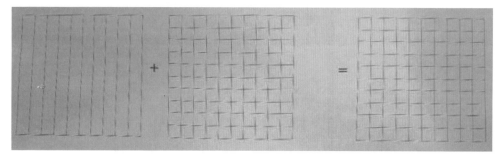

Show the first number, three tenths. Point out the number of parts into whole (10) and the numbers that will be used (3).

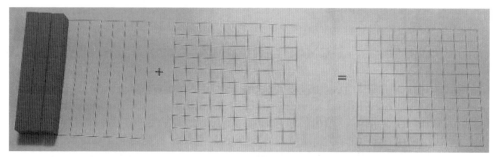

Figure 9–13. Adding Unlike Fractions to Develop Decimals. continues

Show the second number, four hundredths. Point out the number of parts into whole (100) and the parts that will be used (4).

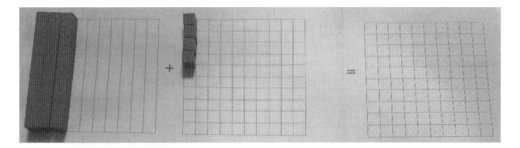

Discuss addition of fractions and the need for both fractions to have the same number of unit fractions in each whole. Show that three tenths is the same as thirty hundredths.

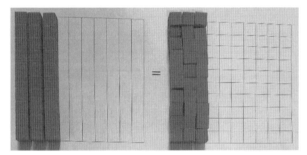

Show the decimal notation for three tenths and thirty hundredths.

0.3 = 0.30

Combine thirty hundredths and four hundredths using the concrete objects. Write the answer using a fraction and show the same equation using decimal notation.

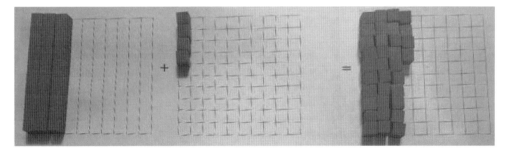

$$\frac{3}{10} + \frac{4}{100} = \frac{34}{100}$$

$$\begin{array}{r} 0.30 \\ +\ 0.04 \\ \hline 0.34 \end{array}$$

Figure 9–13. continues

Representational Level

Begin with problem and translate fraction numbers into decimal notation.

$$\frac{2}{10} + \frac{5}{100} = \frac{}{100}$$

0.2 + 0.05 =

Use tables like the ones below to organize drawings.

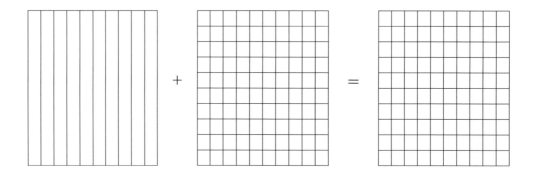

Show the first number, two tenths. Point out the number of parts into whole (10) and the numbers that will be used (2).

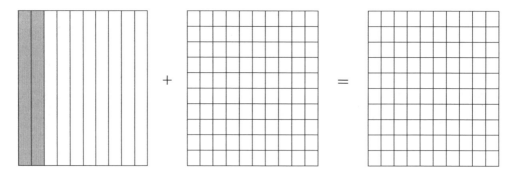

Figure 9–13. continues

Show the second number, five hundredths. Point out the number of parts into whole (100) and the parts that will be used (5).

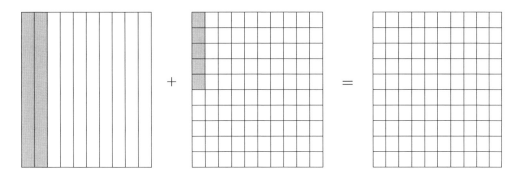

Discuss addition of fractions and the need for both fractions to have the same number of unit fractions in each whole. Show that two tenths is the same as twenty hundredths.

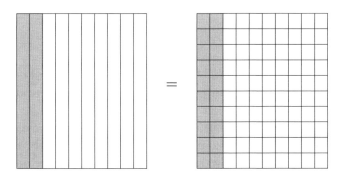

Show the decimal notation for two tenths and twenty hundredths.

0.2 = 0.20

Figure 9–13. continues

Combine twenty hundredths and five hundredths by shading. Write the answer using a fraction and show the same equation using decimal notation.

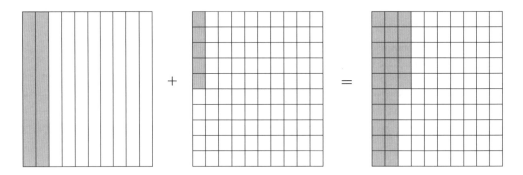

$$\frac{2}{10} + \frac{5}{100} = \frac{25}{100}$$

```
   0. 2 0
+  0. 0 5
─────────
   0. 2 5
```

Figure 9–13. continued

After this knowledge is firm, explore unlike denominators at the concrete and representational/semi-concrete levels and then teach the concepts that define standard procedures that will be used at the abstract level. Finally, the same concepts and teaching sequences that apply to simple operations also apply to beginning understanding of decimals, and CRA/CSA should be used to develop students' conceptual knowledge. CRA/CSA models provide scaffolding and give a conceptual understanding to students so that abstract processes make sense and will be better applied and remembered when students know how and why these are used. Figure 9–14 provides a summary of how CRA/CSA assists students in understanding fractions and operations.

APPLICATION QUESTIONS

1. How are whole-number computational procedures related to computation with fractions?

2. What are common misconceptions for students when computing fractions?

Skill	Common Challenges for Students	How CRA/CSA Intervention Supports Student Learning When These Challenges Are Identified
Adding fractions	Students focus on individual digits and want to put them together, rather than focusing on the meaning of the fraction. For example: 1/3 + 2/5 = 3/8, because students add the top and bottom numbers in error, rather than finding common denominators	Students add the amounts with concrete objects that show students what each fraction represents and assists students in understanding why fractions should have the same denominator and why that denominator does not change when adding and subtracting. Concrete instruction also helps students see that adding fractions is still putting two amounts together, just as it is with whole numbers. This creates a concrete experience that helps students as they transition to the abstract concept.
Adding fractions with different denominators	Students who struggle with multiplication facts and factors may have problems finding common denominators.	Using the area model, particularly using concrete and representational/semi-concrete instruction, provides a strategy for students still working on their factor and multiplication skills.
Multiplication of fractions	Students can struggle when only taught the procedure of multiplying across without context, because it seems too easy and doesn't make sense. Students learn with whole numbers that multiplication is more difficult than addition, so they try to apply this same principle with fractions.	When an interventionist utilizes CRA/CSA, students are able to connect their knowledge of whole-number multiplication to multiplying with fractions (combining groups of same size, now the groups are parts that are the same size such as 1/3 × 1/6 is 1/3 sized groups of each one sixth). This gives context and meaning behind the abstract procedure, which can be helpful when students forget which procedure to use for multiplication. The CSA/CRA described in this chapter begins with not only concrete but also contextual problems to deepen understanding.
Multiplication of fractions	Students believe the product should be larger than the two factors, because multiplication of whole numbers leads to a larger product.	The concrete manipulatives and visuals described in this chapter allow students to see how the basic concept of multiplication is still being utilized, but the product is smaller when two fractions less than 1 are multiplied together. It can also be extended to help students see that two fractions greater than 1 will have a larger product when multiplied together. The deeper conceptual understanding allows students to create generalizations that are more accurate than those at the beginning of the lessons.
Division of fractions	Students believe the quotient should be smaller than the divisor and dividend because of their prior experiences with whole numbers.	By having concrete experiences, such as cutting ribbon using fractional measurements, students are able to explore and make sense of why the quotient can be smaller, depending upon the divisor and dividend.
Division of fractions	Students do not understand why you "turn the second fraction upside down."	This confusion happens when the procedure is introduced, before students have conceptual understanding of division with fractions. Concrete experiences help students gain meaning. Then they are able to make sense of how the mathematical properties are utilized to divide fractions.

Figure 9–14. Ways CRA/CSA Interventions Support Common Challenges Faced by Students.

3. Using any approach, how would addition of fractions be taught at each of the levels of the CRA sequence?

4. Using any approach, how would multiplication of fractions be taught at each of the levels of the CRA sequence?

5. What is the rationale for using each of the following models when teaching addition of fractions: region and length? How do each of these models benefit students?

REFERENCES

Blöte, A. W., Van der Burg, E., & Klein, A. S. (2001). Students' flexibility in solving two-digit addition and subtraction problems: Instruction effects. *Journal of Educational Psychology, 93*, 627–638.

Byrnes, J. P., & Wasik, B. A. (1991). Role of conceptual knowledge in mathematical procedural learning. *Developmental Psychology, 27*(5), 777–786.

Empson, S. B., & Levi, L. (2011). *Extending children's mathematics: Fractions and decimals*. Portsmouth, NH: Heinemann.

Kamii, C., & Warrington, M. A. (1999). Teaching fractions: Fostering children's own reasoning. In L. Stiff & F. R. Curcio (Eds.), *Developing mathematical reasoning in Grades K–12* (pp. 82–92). Reston, VA: National Council of Teachers of Mathematics.

Kelly, B., Gersten, R., & Carnine, D. (1990). Student error patterns as a function of curriculum design: Teaching fractions to remedial high school students and high school students with learning disabilities. *Journal of Learning Disabilities, 23*(1), 23–29.

Ni, Y., & Zhou, Y. (2005). Teaching and learning fractions and rational numbers: The origins and implications of whole number bias. *Educational Psychologist, 40*(1), 27–52.

Rittle-Johnson, B., & Star, J. R. (2007). Does comparing solution methods facilitate conceptual and procedural knowledge? An experimental study on learning to solve equations. *Journal of Educational Psychology, 93*(2), 436–362.

Star, J. R., & Seifert, C. (2006). The development of flexibility in equation solving. *Contemporary Educational Psychology, 31*, 280–300.

Index

Note: Page numbers in **bold** reference non-text material

A

Abstract level
  addition instruction at
    counting on, 71–72
    missing addend, 72
    regrouping, 84, **85**
    RENAME strategy for regrouping, 84, **85**
    subtraction level after, 96
    traditional, 69
  cardinality at, 39
  counting instruction at
    graphic counting, **41**, 42–43
    one-to-one correspondence, 39
    quantities larger than 20, 53, **55**
    shortened counting, 44
    skip counting, 47, **49**
  description of, 6
  fraction computations at, 237, 254–255
  fraction instruction at, 203, 220–221
  multiplication instruction at
    description of, 147
    grouping objects, 135
    grouping objects using arrays, 132
  number sense instruction at
    mathematical stories, 25
    more, less, and same exercises with symbols, 22
  one-to-one correspondence at, 39
  subtraction with regrouping at, 112–114, **114**

Abstraction principle, of counting, 30
Acoustic counting, 30–32, **31**
Addition
  basic, 63, 66
  commutative property in, 61, 69
  conceptual understanding of, 60–61
  description of, 60
  identity property in, 61
  mathematical standards for, 59, **60**
  prerequisite skills for, 60–62
  progress monitoring with, 90–92
  with regrouping, CRA/CSA application for
    at concrete level, 77, **77–81**, 84, **86–90**
    description of, 74–84
    partial sums, 77, **77–79**
    progress monitoring for, 92
    RENAME strategy, 84, **85**
    at representational level, 77, **77–79, 82–83, 86–90**
  rote counting in, 61
  single-digit, 90–91
  symbols used in, 61, 66
Addition instruction
  at abstract level
    counting on, 71–72
    missing addend, 72
    regrouping, 84, **85**
    RENAME strategy for regrouping, 84, **85**
    subtraction level after, 96
    traditional, 69

Addition instruction *(continued)*
    at concrete level
        alternative facts, 73, **75–76**
        counting on, 70, **70**
        missing addend, 72, **73–74**
        regrouping, 77, **77–81**, 84, **86–90**
        subtraction level after, 96
        traditional, 66–68, **67**
    CRA/CSA application for
        alternative facts, 72–74, **75–76**
        basic addition, 63, 66
        counting on, 69–72, **70–72**
        making numbers, 62–63, **64–65**
        missing addend, 72, **73–74**
        regrouping, 74–90, **77–90**
        summary of, 92–93
        traditional instruction, 66–69
    DRAW strategy for, 69
    fractions
        at abstract level, 237
        area model for, 228–231, **229**
        at concrete level, 227–231, **229**
        CRA/CSA application for, 227–231, **229**, 265
        length model for, 231, **232**
        manipulatives for, 230
        number lines for, 231, **232**, 235
        at representational level, 231–235, **234**
        with unlike denominators, 230, 234–235, 237, 259, **260–265**
        without common denominators, 230, 234, **265**
    in kindergarten, 59
    manipulatives for, 63, **67–68**, **70–71**, **73–74**, 79
    numbers and, 61–62
    objects in, 61
    place value in, 61–62
    progress monitoring in, 90–92
    at representational level
        alternative facts, 73–74, **75–76**
        counting on, 71, **71**
        missing addend, 72, **73–74**
        regrouping, 77, **77–79**, **82–83**
        subtraction level after, 96
        traditional, 68, **68**
    sequence of, 59
    traditional, 66–69
Advance organizer, 100
Alternative facts
    for addition instruction, 72–74, **75–76**
    for subtraction instruction, 103–105, **106–107**
Alternative numbers, for addition with regrouping, 84, **86–90**
Analogue quality code, 14
Area model
    for fraction instruction
        comparing and ordering fractions, 204–207, **205–206**, 212–216, **213–215**
        at concrete level, 188–191, **188–191**
        description of, 186
        at representational level, 196, 200, **200**
    for multiplication instruction, 130, **130**, 149, **151–152**, 162
    visuals used with, 188, **188**
Arrays, multiplication instruction by grouping objects using, 130–132, **130–132**
Asynchronous counting, 30, 32, **32**
Automaticity, 71–72

B

Balance scale, **20–22**
Base 10 system
    in addition instruction, 61–62
    in subtraction instruction, 98

C

Cardinal principle, of counting, 30
Cardinality
    at abstract phase/level, 39
    at concrete phase, 36–37, **37**
    by kindergartners, 30
    at representational phase/level, **38**, 38–39
Central executive system, 2
Chanters, 30–31
Circle(s)
    area model use of, 186, 188, 196, 212
    for multiplication instruction, 134

Circle fraction, 228
Codes, 14
Common denominators
  for comparing and ordering fractions, 203–204, 213, 220
  fractions without, computations for, 230
Commutative property, 61, 69
Comparing and ordering fractions
  at abstract level, 220–221
  area model for, 204–207, **205–206**, 212–216, **213–215**
  benchmark numbers for, 218
  common denominators used for, 203–204, 213, 220
  at concrete level, 204–209, **205–206**, **209–211**
  CRA/CSA application, 203–221, **205–206**, **209–211**, **213–215**, **219**
  length model, 208–209, **209–211**, 216–220
  manipulatives for, 203, 205
  number lines for, 217–218, **219**
  overview of, 203–204
  rectangular area model for, 212–213, **215**
  at representational level, 212, **213–215**
  set model for, 207–208, 216, **219**
  visual representations for, 203
Conceptual knowledge
  in addition understanding, 91
  description of, 18
  in subtraction understanding, 112, 124
Concrete level
  addition instruction at
    alternative facts, 73, **75–76**
    counting on, 70, **70**
    making numbers, **64**
    missing addend, 72, **73–74**
    partial sums, 77, **77–79**
    regrouping, 77, **77–81**, 84, **86–90**
    subtraction level after, 96
    traditional, 66–68, **67**
  cardinality at, 36–37, **37**
  counting instruction at
    graphic counting, 40, **41**
    one-to-one correspondence, 36–37, **37**
    quantities larger than 20, 50, **51**
    shortened counting, **43**, 43–44
  fraction computations at
    addition, 227–231, **229**
    multiplication, 249–252, **250**, **252**
    subtraction, 227–231, **229**
  fraction instruction at
    area model, 188–191, **188–191**
    comparing and ordering fractions, 204–209, **205–206**, **209–211**
    length model, 194–196, **197–199**
    set model, 191–192, **192–194**, 194
  multiplication instruction at
    contextual problems, 133–134, **134**
    grouping objects, **133**, 133–134
    grouping objects using arrays, 131, **131**
    number lines, 136, **137**
    partial product approach, 138, **140–142**, 149, **150**
    partial products, 149, **150**
    regrouping, **144–145**
    traditional algorithm, 142, **144–145**
    two-digit multipliers, 149, **150**, **155–157**
  number sense instruction at
    mathematical stories, 23–24
    more, less, and same exercises with objects, 19, **20–21**
    sorting and classifying with objects, 23–24
    one-to-one correspondence at, 36–37, **37**
  subtraction instruction at
    alternative facts, 103–105, **106–107**
    alternative problem approach, **118–122**
    counting up, 103, **104**, **115–117**
    regrouping, 107–108, **108–110**
    summary of, 124
    traditional, 100–101, **101**
Concrete-representational/semi-concrete–abstract sequence. *See* CRA/CSA
Conservation of numbers, 29
Corresponders, 32

Counters, 30, 32
Counting
    abstraction principle of, 30
    acoustic, 30–32, **31**
    asynchronous, 30, 32, **32**
    cardinal principle of, 30
    children's rhymes that foster, **31**
    CRA/CSA for instruction and learning in
        cardinality, 36–39
        counting quantities larger than 20, 50–54, **51–52**, **54**
        developmentally appropriate practice, 33–36
        explicit instruction, 35
        graphic counting, 39–43
        one-to-one correspondence, 36–39
        shortened counting, 39–44
        skip counting, 44–47, 44–49, **46**, **48–49**
        summary of, 54
    development of, 30–33
    graphic. *See* Graphic counting
    irrelevance principle of, 30, 33
    by kindergartners, 30
    one-to-one correspondence principle of
        at abstract phase/level, 39
        at concrete phase/level, 36–37, **37**
        description of, 29–30, 32
        at representational phase/level, **38**, 38–39
    principles of, 29–30
    quantities larger than 20, 50–54, **51–52**, **54**
    resultative, 30, 32–33, **33**
    rote, 61, 91, 122
    sequential development of, 30–33
    shortened. *See* Shortened counting
    skip. *See* Skip counting
    stable order principle of, 30
    stages of, 30–33, **32–34**
    synchronous, 30, 32, **32**
Counting on, 69–72, **70–72**
Counting up, 103, **104–105**, **115–117**
CRA/CSA
    addition instruction using. *See* Addition instruction, CRA/CSA application for
    cardinality using. *See* Cardinality
    counting instruction using. *See* Counting, CRA/CSA for instruction and learning in
    definition of, 3–5
    division instruction using. *See* Division instruction, CRA/CSA application for
    explicit instruction using, 6–7, 35
    fraction computations using. *See* Fraction computations, CRA/CSA application for
    fraction instruction using. *See* Fraction instruction, CRA/CSA application for
    multiplication instruction using. *See* Multiplication instruction, CRA/CSA application for
    number sense instruction using. *See* Number sense instruction, CRA/CSA application for
    subtraction instruction using. *See* Subtraction instruction, CRA/CSA application for

D

DAP. *See* Developmentally appropriate practice
Decimals, 256–259, **257–259**
Declarative knowledge
    in addition understanding, 91
    description of, 18
    in subtraction understanding, 98
Developmentally appropriate practice
    core considerations of, 34
    for counting instruction, 33–36
    definition of, 33
    teaching strategies for, 33, 35
Division
    conceptual understanding of, 167
    description of, 168–169
    large numbers, 176–180, **178**
    mathematical standards for, 167–168, **169**
    prerequisite skills for, 168–169
    with remainders, 173–175, **174–175**
Division instruction
    at abstract level, 173

at concrete level
    equal groups for teaching division, 170–171, **171**
    large numbers, 176–177, **178**
    with remainders, 173, **174**
    traditional algorithm, 178–180, **180**
CRA/CSA application for
    basic division, 169–173
    equal groups for teaching division, **170–171**, 170–173
    large numbers, 176–180, **178**
    with remainders, 173–175, **174–175**
    traditional algorithm, 178–180, **180**
DRAW strategy for, 173
fractions, 255–256, **265**
manipulatives for, 170
partial quotients, 177–178, **179**
problem solving, 175, **176**
at representational level
    equal groups for teaching division, 171–173, **172**
    with remainders, 174, **175**
sequence of, 167–168
summary of, 181
traditional algorithm for, 178–180, **180**
DRAW strategy
for addition instruction, 69
for division instruction, 173
for multiplication instruction, 135
Drawings
addition instruction using, 68, **68**
division instruction using, 171, 174, **175**
mathematical stories using, 24–25
more, less, and same exercises with, 19–21, **22**
multiplication instruction using, 134
subtraction instruction using, **102**, 102–103, **105**

E

Enactive stage, 3, 12
Episodic buffer, 2
Equal sign, 66, 72, 91

Equations
    chart of, 254–255, **255**
    skip counting with, **55**
Equipartitioning
    of fractions, 185, 202, 217, 226
    of isosceles triangle, 186
Equivalent decimals, 256, **257–259**
Equivalent fractions
    at abstract level, 220
    area model for, **238–241**, **244–247**
    at concrete level, 204, **238–243**
    equivalent decimals from, 256, **257–259**
    length model for, **242–243**, **248–249**
    number line for finding, 218–220
    overview of, 203–204
    at representational level, 212, **213–214**, **244–249**
Evidence-based practices, 35
Explicit instruction
    description of, 6–7, 35, 67
    in fraction instruction, **189–190**, **193**
    in subtraction, 95, 100
    with unit fractions, **198–199**
Expressive language, 5

F

Fact families
    for division, 168
    for subtraction, 97
Factors, 129
Feedback, 21, 43
Flashcards, graphic counting using, 40, **41**
Flexible counting, 30–31
Fluency, 71, 97
Fraction(s)
    benchmark numbers compared to, 218
    composition of, 196
    concept of, 188, 192
    description of, 184–186
    equipartitioning in, 185, 202, 217, 226
    meanings of, 184, 226
    overview of, 183–184
    prerequisite skills for, 184–186

Fraction(s) *(continued)*
  unit
    addition of, 196, **198–199**
    definition of, 185
    description of, 185–186, 225
    division of whole number by, 256, **256**
    explicit instruction involving, **198–199**
    unitizing and, 185
    as whole, 185, 188, **209–211**
Fraction bars
  for addition of fractions, 231
  for comparing and ordering fractions, 208
  length model's use of, 187, 194–195, 208
  for multiplication of fractions, 251–252
  for subtraction of fractions, 231
Fraction computations
  at abstract level, 237, 254–255
  addition
    at abstract level, 237
    area model for, 228–231, **229**
    at concrete level, 227–231, **229**
    CRA/CSA application for, 227–231, **229**, **265**
    length model for, 231, **232**
    manipulatives for, 230
    number lines for, 231, **232**, 235
    at representational level, 231–235, **234**
    with unlike denominators, 230, 234–235, 237, 259, **260–265**
    without common denominators, 230, 234, **265**
  area model for
    addition, 228–231, **229**, 234–235
    at concrete level, 228–231, **229**
    multiplication, 251, **252**, 253
    at representational level, 234–235, **236**
    subtraction, 228–231, **229**, 234–235
  at concrete level
    addition, 227–231, **229**
    multiplication, 249–252, **250**, **252**
    subtraction, 227–231, **229**
  CRA/CSA application for
    addition, 227–249, **229–249**, **265**
    division, 255–256, **265**
    multiplication, 249–255, **265**
    subtraction, 227–249, **229–249**, **265**
    summary of, **265**
  decimals, 256–259, **257–259**
  description of, 226–227
  division, 255–256, **265**
  length model for
    addition, 235
    multiplication, 251–254
    at representational level, 235
    subtraction, 235
  multiplication
    at abstract level, 254–255
    area model for, 251, **252**, 253
    chart of equations, 254–255, **255**
    at concrete level, 249–252, **250**, **252**
    CRA/CSA application for, 249–255, **265**
    length model for, 251–254
    model for, 250, **250**
    at representational level, 253–254, **254**
    whole-number, 249–250
  overview of, 225
  prerequisite skills for, 226–227
  at representational level
    addition, 231–235, **234**
    area model for, 234–235, **236**
    length model for, 235
    multiplication, 253–254, **254**
    subtraction, 231–235, **234**
  sequence of instruction for, 225–226
  subtraction
    at abstract level, 237
    area model for, 228–231, **229**
    at concrete level, 227–231, **229**
    CRA/CSA application for, 227–231, **229**, **265**
    with different denominators, 230
    length model for, 231, **232**
    manipulatives for, 230
    at representational level, 231–235, **234**

summary of, 259, 264
with unlike denominators, 230, 234–235, 237, 259, **260–265**
whole-number computation knowledge for, 226
Fraction instruction
at abstract level, 203, 220–221
area model for
comparing and ordering fractions, 204–207, **205–206**, 212–216, **213–215**
at concrete level, 188–191, **188–191**
description of, 186
equivalent fractions, **238–241**, **244–247**
at representational level, 196, 200, **200**
benchmark numbers, 218
in comparing and ordering fractions
at abstract level, 220–221
area model for, 204–207, **205–206**, 212–216, **213–215**
benchmark numbers, 218
common denominators, 203–204, 213, 220
at concrete level, 204–209, **205–206**, **209–211**
CRA/CSA application, 203–221, **205–206**, **209–211**, **213–215**, **219**
length model, 208–209, **209–211**, 216–220
manipulatives for, 203, 205
number lines for, 217–218, **219**
overview of, 203–204
rectangular area model for, 212–213, **215**
at representational level, 212, **213–215**
set model for, 207–208, 216, **219**
visual representations for, 203
in computation of fractions. *See* Fraction computations
at concrete level
area model, 188–191, **188–191**
comparing and ordering fractions, 204–209, **205–206**, **209–211**

length model, 194–196, **197–199**
set model, 191–192, **192–194**, 194
CRA/CSA application for
at abstract level, 203
comparing and ordering fractions, 203–221, **205–206**, **209–211**, **213–215**, **219**
at concrete level, 187–196, **188–195**, **196–199**
fraction computations. *See* Fraction computations, CRA/CSA application for
at representational level, 196–202, **200–202**
summary of, 221
equivalent fractions
at abstract level, 220
area model for, **238–241**, **244–247**
at concrete level, 204, **238–243**
equivalent decimals from, 256, **257–259**
length model for, **242–243**, **248–249**
number line for finding, 218–220
overview of, 203–204
at representational level, 212, **213–214**, **244–249**
explicit instruction in, **189–190**, **193**
in fraction computations. *See* Fraction computations
length model for
in comparing and ordering fractions, 208–209, **209–211**, 216–220, **219**
at concrete level, 194–196, **197–199**
description of, 187
equivalent fractions, **242–243**, **248–249**
fraction bars used in, 187, 194–195
number line as, 194–195, **197**, 202, **202**
at representational level, 201–202, **202**, 216–220
manipulatives for, 187, 191–192, 196, 203–206, 208, 212, 216, 227–228, 230, 251–253
manipulatives used in, 191

Fraction instruction *(continued)*
   measurement model for, 195
   at representational level
     area model, 196, 200, **200**
     comparing and ordering fractions, 212, **213–215**
     equivalent fractions, 212, **213–214**, **244–249**
     length model, 201–202, **202**, 216–220
     set model, 200–201, **201**, 216
   sequence of, 184
   set model for
     in comparing and ordering fractions, 207–208, 216
     at concrete level, 191–192, **192–194**, 194
     description of, 187
     at representational level, 200–201, **201**, 216
   summary of, 221
   visual models used in, 186–187
Fractional meanings, 184, 226

G

Graphic counting
   at abstract level, **41**, 42–43
   at concrete level, 40, **41**
   definition of, 39
   at representational level, 40–42, **41–42**
Guided practice, 6

H

Hundreds chart, for skip counting, 53, **54–55**

I

Iconic stage, 3, 12
Identity property, 61, 220
Information-processing deficits, 5
Irrelevance principle, of counting, 30, 33
Isosceles triangle, 186

J

Joining. *See* Addition
Journals, 24

K

Kindergarten
   addition instruction in, 59
   cardinality in, 30
   counting in, 30
   number concept understanding in, 2
   number sense understanding in, 11
   subtraction instruction in, 95

L

Language development, 25
Language processing difficulties, 5
Large numbers, division of, 176–180, **178**
Length model, for fraction instruction
   addition of fractions, 231, **232**
   in comparing and ordering fractions, 208–209, **209–211**, 216–220, **219**
   at concrete level, 194–196, **197–199**
   description of, 187
   fraction bars used in, 187, 194–195
   number line as, 194–195, **197**, 202, **202**
   at representational level, 201–202, **202**, 216–220
   subtraction of fractions, 231, **232**
Like denominators, 212
Like numerators, 212

M

Making numbers, 62–63, **64–65**
Manipulatives
   addition instruction using, 63, **67–68**, **70–71**, **73–74**, 79
   division instruction using, 170
   fraction instruction using, 187, 191–192, 196, 203–206, 208, 212, 216, 227–228, 230, 251–253
   multiplication instruction using, 133–134, 136, 138, 141–142, **265**

subtraction instruction using, 101–102
unit fraction instruction using, 186
virtual, 5
Mapping skills, in number sense instruction, 14–16, **15–17**
Mathematical standards
for addition, 59, **60**
for division, 167–168, **169**
for multiplication, 127–129, **128**, 163
for subtraction, 95–96, **96**
Mathematical stories
at concrete level, 23–24
definition of, 23
drawings used in, 24–25
read-alouds and, 23
sorting and classifying with objects, 23–24
symbols used in, 25
Mathematics learning disability, 2
Measurement model, for fraction instruction, 195
Memory
language development and, 25
working, 2
Mental representation, Bruner's stages of, 3
Minuend, 102–103
Multiplicands, 129, 153–154
Multiplication
description of, 128–129
mathematical standards for, 127–129, **128**
prerequisite skills for, 128–129
progress monitoring in, 161–162
single-digit, 161–162
whole-number, 249–250
Multiplication instruction
at abstract level, 147
grouping objects, 135
grouping objects using arrays, 132
area model for, 130, **130**, 149, **151–152**, 162
arrays
grouping objects using, 130–132, **130–132**
partial products and, 138–140
at concrete level

contextual problems, 133–134, **134**
grouping objects using arrays, 131, **131**
number lines, 136, **137**
partial product approach, 138, **140–142**, 149, **150**
regrouping, **144–145**
traditional algorithm, 142, **144–145**, 153
two-digit multipliers, 149, **150**, **155–157**
CRA/CSA application for
grouping objects, 133–136
grouping objects using arrays, 130–132, **130–132**
one-digit multipliers, 136–147, **137–148**, 162
summary of, 163–164
traditional algorithm, 149–160, **152–161**
two-digit multipliers, 148–160, **149–161**
double-digit, 162
fractions
at abstract level, 254–255
area model for, 251, **252**, 253
chart of equations, 254–255, **255**
at concrete level, 249–252, **250**, **252**
CRA/CSA application for, 249–255, **265**
length model for, 251–254
model for, 250, **250**
at representational level, 253–254, **254**
whole-number, 249–250
manipulatives for, 133–134, 136, 138, 141–142, **265**
number sense in, 148, 154
one-digit multipliers
description of, 136–138
partial products, 138, **139**
progress monitoring with, 162
partial products approach
area models, 149, **151–152**
one-digit multipliers, 138–141, **139**
two-digit multipliers, 149, **150–151**
place value mat in, 149, 153–154

Multiplication instruction *(continued)*
  progress monitoring during, 161–162
  regrouping, 162
  with regrouping, **144–147**
  RENAME strategy, 147, **148**, 154–155
  at representational level
    grouping objects, **133**, 133–134, **135**
    grouping objects using arrays, 132, **132**
    number lines, 136, **137**
    partial product approach, 138, **140–142**, 149, **151**
    regrouping, **146–147**
    traditional algorithm, **146–147**
    two-digit multipliers, **151**, 154, **155**, **158–161**
  sequence of, 127–128
  summary of, 163–164
  traditional algorithm
    for one-digit multipliers, **146–147**
    for two-digit multipliers, 149, 151–161
  two-digit multipliers, 148–160, **149–161**
Multiplication mat, **139**

N

Nonsymbolic information
  coding of, 15
  mapping of, **17**
Nonsymbolic problem solving, **18**
Nonsymbolic skills
  CRA/CSA sequence for
    application of, 18–22, **20–22**
    description of, 17–18
    more, less, same exercises, 19–22, **20–22**
  examples of, 15, **16**
Number(s)
  in addition instruction, 61–62
  base 10 system for, 61–62
  conservation of, 29
  making of, 62–63, **64–65**
  place value of, 61–62
  prerequisite skills for, 61
  skip counting using, 47, **49**

Number line
  for counting on, 71, **72**
  equivalent fractions, 218–220
  errors associated with, 71
  fraction instruction using
    addition of fractions, 231, **232**, 235
    comparing and ordering fractions, 208, 217, **219**
    at concrete level, 194–195, **197**
    multiplication of fractions, 253–254
    at representational level, 202, **202**
  multiplication instruction using
    fractions, 253–254
    whole numbers, 136, **137**
  partitioning of, 218
Number sense instruction
  at abstract level
    mathematical stories, 25
    more, less, and same exercises with symbols, 22
  cognitive underpinning of, 14
  at concrete level
    mathematical stories, 23–24
    more, less, and same exercises with objects, 19, **20–21**
    sorting and classifying with objects, 23–24
  counting. *See* Counting
  CRA/CSA application for
    more, less, and same exercises, 18–22, **20–22**
    nonsymbolic skills, 18–22, **20–22**
    sorting and classifying exercises, 23–25
  definition of, 11
  description of, 11–12
  development of, 14, 25
  elements of, 12
  mapping skills, 14–16, **15–17**
  mathematical representations and, 12, **13**
  in multiplication instruction, 148, 154
  nonsymbolic skills. *See* Nonsymbolic skills
  quantity knowledge, 30
  reasoning in, 63

at representational level
  mathematical stories using drawings, 24–25
  more, less, and same exercises with drawings, 19–21, **22**
Number symbols
  mapping skills with, 14–16, **15–17**
  quantities and, making connections between, 14–16, **15–17**, 30
Numeracy. *See* Number sense
Numeric working memory, 2
Numerical tasks, 2

O

Objects
  in addition instruction, 61
  basic addition using, 66
  counting quantities larger than 20 using, 50, **51**
  more, less, and same exercises with, 19, **20–21**
  sorting and classifying with, 23–24
One rule, 69
One-to-one correspondence
  at abstract phase/level, 39
  in addition, 61
  at concrete phase/level, 36–37, **37**
  description of, 29–30, 32
  at representational phase/level, **38**, 38–39
Order rule, 69
Ordering fractions. *See* Comparing and ordering fractions

P

Partial products approach
  area models, 149, **151–152**
  one-digit multipliers, 138–141, **139**
  two-digit multipliers, 149, **150–151**
Partial quotients, for large-number division instruction, 177–178, **179**
Phonological loop, 2
Pictures
  basic addition using, 66
  counting quantities larger than 20 using, 50–51, **52**
  skip counting using, 47
Place value
  in addition instruction, 61–62, 79
  in subtraction instruction, 98
Place value mats
  for addition instruction, 79
  for multiplication instruction, 143, 149, 153–154
  for subtraction with regrouping instruction, 106–107, 111, 115
Place value table, 141, **142**
Plus sign, 91
Problem solving
  addition instruction using, 59, 66, 92
  division instruction using, 173–175, **176**, 180
  multiplication instruction using, 127, 129, 131, 133, 141, 163
  nonsymbolic, **18**
  partial products used for, **150**
  RENAME strategy for, **148**
  Standards for Mathematical Practice, 59
  subtraction instruction using, 100–101, 107
  symbolic, **18**
Procedural knowledge
  in addition understanding, 92
  description of, 18
  in subtraction understanding, 112
Producers, 30, 32

Q

Quantities
  number symbols and, making connections between, 14–16, **15–17**, 30
  student awareness of, 96
Quantity knowledge, in number sense, 30

R

Read-alouds, 23
Reasoning, 63
Receptive language, 5
Reciters, 30, 32

Rectangle fraction, 212
Rectangles
   area model use of, 188, **188**
   for fraction instruction, 213, 235
Regrouping
   addition with
      at concrete level, 77, **77–81**, 84, **86–90**
      description of, 74–84
      progress monitoring for, 92
      at representational level, 77, **77–79**, **82–83**, **86–90**
   subtraction with
      at abstract level, 112–114, **114**
      alternative problem strategy, 115, **118–122**
      at concrete level, 107–108, **108–110**
      description of, 106–122
      place value mats, 106–107
      progress monitoring, 123
      RENAME strategy, 112–114, **114**
      at representational level, 111, **111–113**
      research regarding, 106–107
RENAME strategy
   for addition with regrouping, 84, **85**
   for multiplication with regrouping, 147, **148**, 154–155
   for subtraction with regrouping, 112–114, **114**
Representational, 4
Representational level
   addition instruction at
      alternative facts, 73–74, **75–76**
      counting on, 71, **71**
      making numbers, **65**
      missing addend, 72, **73–74**
      partial sums, 77, **77–79**
      regrouping, 77, **77–79**, **82–83**
      subtraction level after, 96
      traditional, 68, **68**
   cardinality at, **38**, 38–39
   counting instruction at
      graphic counting, 40–42, **41–42**, **42**
      quantities larger than 20, 50–53, **52**
      shortened counting, 44, **45**
      skip counting, 47, **48**
   fraction computations at
      addition, 231–235, **234**
      area model for, 234–235, **236**
      length model for, 235
      multiplication, 253–254, **254**
      subtraction, 231–235, **234**
   fraction instruction at
      area model, 196, 200, **200**
      comparing and ordering fractions, 212, **213–215**
      equivalent fractions, 212, **213–214**, **244–249**
      length model, 201–202, **202**, 216–220
      set model, 200–201, **201**, 216
   multiplication instruction at
      grouping objects, **133**, 133–134, **135**
      grouping objects using arrays, 132, **132**
      number lines, 136, **137**
      partial product approach, 138, **140–142**, 149, **151**
      regrouping, **146–147**
      traditional algorithm, **146–147**
      two-digit multipliers, **151**, 154, **155**, **158–161**
   number line. *See* Number line
   number sense instruction at
      mathematical stories using drawings, 24–25
      more, less, and same exercises with drawings, 19–21, **22**
      one-to-one correspondence at, **38**, 38–39
   subtraction instruction at
      alternative problem approach, **118–122**
      counting up, 103, **105**, **115–117**
      regrouping, 111, **111–113**
      summary of, 124
      traditional, 102, **102**
Representations
   number sense and, 12, **13**
   stages of, 3, 12, **13**
   types of, 3
Resultative counting
   description of, 30, 32–33, **33**
   graphic counting with, 43

Rhymes, for counting, **31**
Rote counting, 61, 91, 122

S

Set model, for fraction instruction
   in comparing and ordering fractions, 207–208, 216
   at concrete level, 191–192, **192–194**, 194
   description of, 187
   at representational level, 200–201, **201**, 216
Shortened counting
   at abstract level, 44
   at concrete level, **43**, 43–44
   CRA/CSA application for, 39–44
   description of, 30, 33
   equation for, **55**
   example of, **34**
   at representational level, 44, **45**
Skip counting
   at abstract level, 47, **49**
   at concrete level, 44–46, **46**
   in counting quantities larger than 20, 51, 53
   equation for, **55**
   hundreds chart for, 53, **54–55**
   modeling of, 53
   reasons for, 51, 53
   at representational level, 47, **48**
   ten frames for, 44–45, **46**
Stable order principle, of counting, 30
Standards for Mathematical Practice
   for addition, 59, **60**
   description of, 59
Subitizing. *See* Graphic counting
Subtraction
   definition of, 96
   description of, 96–98
   mathematical standards for, 95–96, **96**, 163
   prerequisite skills for, 96–98
   progress monitoring in, 122–123
   with regrouping, CRA/CSA application for
     at abstract level, 112–114, **114**
     alternative problem strategy, 115, **118–122**
     at concrete level, 107–108, **108–110**
     description of, 106–122
     place value mats, 106–107
     progress monitoring, 123
     RENAME strategy, 112–114, **114**
     at representational level, 111, **111–113**
     research regarding, 106–107
   single-digit, 122–123
   student errors in, **99**
Subtraction instruction
   at concrete level
     alternative facts, 103–105, **106–107**
     alternative problem approach, **118–122**
     counting up, 103, **104**, **115–117**
     summary of, 124
     traditional instruction, 100–101, **101**
   CRA/CSA application for
     alternative facts, 103–105, **106–107**
     counting up, 103, **104–105**, **115–117**
     explicit instruction, 100
     overview of, 98–100
     summary of, 124
     traditional instruction, 100–102, **101–102**
   fractions
     at abstract level, 237
     area model for, 228–231, **229**
     at concrete level, 227–231, **229**
     CRA/CSA application for, 227–231, **229**, 265
     with different denominators, 230
     length model for, 231, **232**
     manipulatives for, 230
     at representational level, 231–235, **234**
   in kindergarten, 95
   manipulatives for, **101–102**
   progress monitoring during, 122–123
   at representational level
     alternative problem approach, **118–122**
     counting up, 103, **105**, **115–117**
     summary of, 124
     traditional instruction, 102, **102**
   sequence of, 95–96
Subtrahend, 102–103

Symbol(s)
   mathematical, 61, 66
   mathematical stories using, 25
Symbolic information
   coding of, 15
   mapping of, 14–15, **17**
Symbolic numbers, **16**
Symbolic problem solving, **18**
Symbolic representation stage, 3, 12
Symbolic skills
   CRA/CSA sequence used to teach
      application of, 18–22, **20–22**
      description of, 17–18
      more, less, same exercises, 19–22, **20–22**
   examples of, **16**
   identifying of, 15
Synchronous counting, 30, 32, **32**

T

Ten frames
   counting pictures with, 40, 42, **42**
   skip counting using, 44–45, **46**

U

Unit fractions
   addition of, 196, **198–199**
   definition of, 185
   description of, 185–186, 225
   division of whole number by, 256, **256**
   explicit instruction involving, **198–199**
Unitizing, 185
Unlike denominators, fraction computations with, 230, 234–235, 237, 259, **260–264**

V

Verbal code, 14
Virtual manipulatives, 5
Visual code, 14
Visual-spatial sketchpad, 2

W

Word problems
   in addition instruction, 93
   equations created from, 66, 68
   in subtraction instruction, 100
Working memory, 2

Z

Zero rule, 69